ORIGINS
OF LIFE

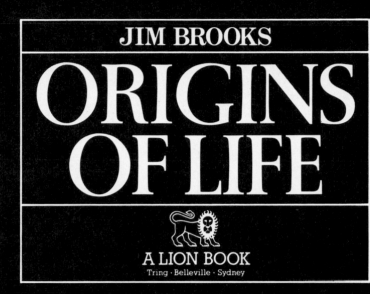

JIM BROOKS

ORIGINS OF LIFE

A LION BOOK
Tring · Belleville · Sydney

Copyright © 1985 Dr Jim Brooks

Published by
Lion Publishing plc
Icknield Way, Tring, Herts, England
ISBN 0 85648 809 7
Lion Publishing Corporation
10885 Textile Road, Belleville, Michigan 48111, USA
ISBN 0 85648 809 7
Albatross Books
PO Box 320, Sutherland, NSW 2232, Australia
ISBN 0 86760 628 2

First edition 1985

Printed in Hong Kong

CONTENTS

TODAY'S ANSWERS TO THE AGE-OLD QUESTIONS

The scope of this book is vast. I begin with the origin of the Universe and its age. Then we see how galaxies and stars were formed, how our Solar System came to be, and our own Planet Earth. We study the nature of the early Earth, and ask how life could have begun on it. So the range is wide, and the issues are fascinating.

I have tried not to be overdogmatic, always making it clear when the answers to questions raised in the story are not yet known for sure. And more is known now than a few years ago. The scientific environment from the late 1960s through to the 1980s was made for studies of the origins of life.

Of course people have always been interested in these questions. But the time is ripe to study them. There comes a time when a subject becomes ripe for scientific investigation. Until that particular time comes, no matter how bright or clever scientists may be, advances are unlikely. Answers to specific scientific problems frequently require methodology and knowledge without which the problems must remain unsolved.

During these last three decades a number of different areas of science have reached the point where such methods and knowledge have begun to be attained:

● The basic discoveries of **chemistry, physics and biology** have come together in a quite remarkable way to give an insight into the nature of the living system;

● Significant discoveries have been made in **cosmology**;

● Spectacular achievements in **space technology** have produced results which could not even have been attempted or imagined a few decades ago, including unmanned craft visiting other planets;

● **Microanalytical techniques**, so useful to biologists, have also now given the tools to identify trace amounts of material, which have helped the **geologist, geochemist and cosmochemist** to identify trace fossils in very ancient rocks and extra-terrestrial materials.

Space shuttle Columbia is launched at Cape Kennedy in Florida. Space technology has produced spectacular achievements in recent years, including unmanned spacecraft visiting other planets. All this has led to some exciting scientific discoveries, some of them highly relevant to questions about the origins of life.

There are, however, reasons other than purely scientific ones which bring a sense of excitement to the study of origins. There is an obvious overlap with religious questions. It is no accident that the story I am telling — from the first moments of the Universe to the beginnings of life — is narrated in a quite different way in the first chapter of Genesis.

Many people believe that science and faith are incompatible, and are surprised to learn that scientists, especially those involved in studying the origin of life and the nature of the Universe, are also Christians. But the combination is a good deal more common than usually thought.

The sort of topics in this book — 'The Origin of the Universe', 'Life in Space', 'The Origin of Life' — are especially prone to be treated in a way that dramatizes any alleged conflict between science and religion. This is not my approach. To my mind, science and religion are about different types of knowledge about the same Universe.

This is a book about science, written by a scientist. But these religious questions are not ducked or treated as unimportant. I hope that in this way the whole range of the reader's questions will at least get an airing, and that he or she will come away stimulated and informed.

In the Beginning

Scientists now believe that the Universe began at a specific moment in time with a vast explosion. How can we know what happened billions of years ago? And what is the Universe made of? How were the stars, planets and moons formed? How did life begin on Earth? Has it all happened by chance or is there a design?

In the beginning, there was nothing at all. This is a very difficult concept. It causes a great deal of scientific misunderstanding among many people who have heard of the 'Big Bang' — the creation of the Universe as we know it in a massive explosion of matter and energy.

When we look up at the night sky it gives a powerful impression of a changeless Universe. True, clouds drift by the Moon, the sky rotates around the polar star, and over longer periods the Moon itself waxes and wanes and the Moon and the planets move against the background of stars. These events are mere local phenomena caused by motions within our solar system. Beyond the planets, the stars seem motionless.

How does this fit with the notion of a Big Bang? From our everyday experience, we know what a big bang must be like — a concentration of matter, triggered by some energetic process blasting outward into space. It is a very different picture from the apparently peaceful Universe we contemplate in the night sky.

But in fact those still-looking stars are moving, at speeds ranging up to a few hundred miles per second, so that in the Earth's calendar-year a fast star might travel ten thousand million miles or so. This is a thousand times less than the distance to even the closest stars, so their apparent position in the sky changes very slowly, and can give the impression of being motionless.

Before the Big Bang of creation, there was not even any empty space. Space and Time, as well as energy and matter, were created in this creative explosion. There was no

As we look at this far-distant system of stars, we could almost be looking at ourselves from many light-years away in space. This is a spiral galaxy — a huge disc of stars, gas and dust which is slowly rotating on its axis. Our own galaxy is very similar.

'outside' for the exploding Universe to explode into, since even when it was only just born and beginning its great expansion, the Universe contained everything, including all the empty space.

Mankind has always sought to search out his beginnings. Throughout history kings and princes, even the most primitive tribes, have had their astronomers and before that

THE BIG-BANG THEORY

The Big Bang is undoubtedly the most widely-held theory to account for the formation of galaxies and indeed of the whole Universe. Almost all physical observations made by cosmologists and radio-astronomers have tended to confirm the theory's basic concepts.

The assumption is that somewhere about 10,000 million years ago the Universe consisted of a single piece of matter, a sort of giant primeval atom concentrated at a single point in space. This concentrated matter then exploded, flinging out either hydrogen atoms and related nuclei which subsequently became organized into discrete units (galaxies) or else (or as well as) lumps of specifically-condensed matter which could be regarded as incipient galaxies.

These lumps of matter or more tenuous gas have ever since continued to move away from each other at ever-increasing speed. Presumably, since we can only observe those objects which travel at speeds less than that of light, as the speed approaches that of light the objects will disappear and ultimately the whole galactic Universe will disappear.

Observations seem to confirm these arguments, since at greater distances the concentration of galaxies apparently increases, as one would expect on this model. In addition, there is evidence that the whole Universe is emitting radiation at radio wavelengths resembling a very cold black body at about 3° Kelvin. Such radiation had been predicted as a natural consequence of the primeval Big Bang.

Some cosmologists think that the super-condensed primeval atom came about through an earlier contraction of a rarefied Universe. The assumption is that at a certain degree of rarefaction our own present Universe will begin to contract and ultimately be yet again reborn in another giant explosion. This particular theory, while fitting what facts there are, has nevertheless had an uneasy reception. This is largely because it involves tightly-scheduled starts and finishes in a Universe which somehow many feel to be endless and infinite.

their astrologers. They have looked up into the stars and sought to understand the origin and structure of the Universe. During most of human history these observations in matters of navigation, magic and futurology failed to give much scientific understanding; general theories about the Universe were a very long time being formulated.

It was long recognized from their degrees of brightness that stars were at different distances from the Earth, but measurement of those distances and the discovery of relative stellar motion was slow in coming.

CONTINUOUS-CREATION / STEADY-STATE THEORY

Many people disliked the idea that time had a beginning or an end, because it suggested divine intervention. There have been a number of attempts to avoid this conclusion. One of these was the 'continuous-creation/steady-state' model of the Universe proposed in 1948 by Hermann Bondi, Fred Hoyle and Tom Gold. This model, one suspects, was designed largely to overcome psychological as well as mathematical objections to the Big-Bang theory, and indeed it has been criticized on this basis as a sort of search for perfection in nature.

In the steady-state model it was proposed that as the galaxies moved farther away from each other new galaxies were formed in between, out of matter that was being 'continually created'. The Universe would therefore look more or less the same at all times and its density would be roughly constant. The model suggests that matter (in the form of hydrogen) is always being created from nothing, and that this happens in order to counteract the dilution of material which occurs as the galaxies drift away from each other.

The amount of matter required to be produced on this theory is very small indeed — about one atom of hydrogen per litre per 10^{12} years — although the rate of creation could obviously be much greater in specific pockets. However, it must not be created in too non-random a fashion or it might not do the job intended — to renew lost galactic material. In addition, the newly-created matter will have the same velocity as its immediate neighbours, so that the red-shift effect will not be contravened (see *The Doppler Effect*).

A theory of this sort might seem at first sight to be extreme. However, one can get an idea as to how matter can be created from 'nothing' by invoking the very useful concept of 'ether'. Ether we can say has the property of being 'nothing' in the sense that it cannot be measured, but warped (condensed and rarefied) ether becomes something in that it can be measured. Thus we say that ether has the property that allows it to take up specific discrete, not continuous, small condensed forms (elementary particles). But its total properties are such that, when

such a condensed form appears, a rarefied equivalent piece of ether (elementary anti-particle) must also appear. So from apparently nothing we obtain two particles. (The mathematician, however, claims to produce the same sort of result without invoking an ether. And in addition he provides quantitative data to support his theory.)

This model had the great virtue that it made definite predictions that could be tested by observations. Unfortunately, observations of radio sources by Professor Martin Ryle at the University of Cambridge in the 1950s and 60s showed that the number of radio sources must have been greater in the past, contradicting the steady-state theory. Probably the final scientific observation that disproved this theory was the discovery in 1965 of microwave background radiation in the Universe. There was no way this radiation could be accounted for in the model.

AN EXPANDING UNIVERSE

The key to the science of cosmology today is the discovery that the Universe is expanding. It is now generally believed that the Big-Bang creation began this expansion and saw the beginning of everything.

Twenty years ago cosmologists were divided between those who postulated that the Universe is expanding and originated in the Big Bang, and those who believed in an eternal process of continuous creation. Superficially it appeared that the Continuous-Creation hypothesis dispensed with God, so many theologians favoured the Big-Bang theory. It fitted in conveniently with the opening verse of Genesis 1, 'In the beginning . . .'

It is now known for sure that the Universe is expanding. It is in a state of violent explosion, in which great islands of stars (galaxies) are rushing apart at speeds approaching the speed of light.

Theoretical physicists and astronomers are able to extrapolate this explosion backward in time and conclude that all the galaxies must have been much closer at some time in the past. At these very early times, extended backward to within minutes or seconds of the creation, the matter would have been so close that neither galaxies nor stars nor even atoms and atomic nuclei could have had a separate existence: all would have been fused in an unbelievably dense lump with no spaces at all. This is the era which is called 'the early Universe'.

The hot Big-Bang model provides a good framework for understanding the early evolution and present structure of the Universe. It has proved to be remarkably useful — to ignore its success would be scientifically dishonest. But there are problems. Many questions and ambiguities are neither addressed nor answered within the concept of the standard Big-Bang model. For example, our present Universe is very old (approximately 10,000 million years) and yet very young in the sense that spatial curvature is not yet dominant; the Universe is very 'flat'. One response to this age-flatness puzzle, on the basis of the Big-Bang model of initial conditions, is that 'the Universe is the way it is because it was the way it is'.

Dating the
Distant Past

'In the beginning . . .' brings us immediately into a world of space and time. Space and time are like warp and weft and their interwoven relationship is history. So our attempts to understand the beginning of the Universe, of the Solar System, or of Planet Earth deal in the realm of history, just as much as when we try to understand the walls of Jericho falling down or the American Civil War.

Of course, history is normally thought of as depending on written records. But the distant past has its records too, although, as we shall see, they are of a quite different sort.

A sense of time appears to be one of the marks which distinguishes mankind from the other animals. Primitive and ancient peoples show by their burial customs a respect for the past and a concern for the future. And modern, civilized men and women put a great deal of effort into trying to understand and reconstruct the past. Many of us today tend to escape into the past, on our regular pilgrimages to ancient cathedrals, castles and battle-grounds. We are fascinated by the records and remains of events of tens, hundreds and sometimes thousands of years ago.

With the recent books and television programmes in the area of popular science, many of us have been introduced to events and environments that occurred, not just a thousand or a million years ago, but sometimes hundreds and thousands of million years ago. In these vast 'eons' of time,

Overleaf
The uniquely impressive Grand Canyon, a mile-deep gorge in Arizona, was formed as rocks were cut away over millions of years by the Colorado river. The rocks still lie roughly horizontal, making a kind of natural Geological Column. The rocks at the bottom of the canyon are 2,000 million years old.

This fossil megaspore, Valvisisporites auritus, **is derived from an ancient plant which lived about 250 million years ago in the Carboniferous Age. It was found in the Lawrence Shale of the Lone Starr Lake region, Kansas, United States.**

THE CALENDAR OF EARTH'S HISTORY

If we imagine Earth's history as a single calendar-year, it will look like this. Each day represents about 12 million years. So much of life's development has taken place in the last 550 million years, that we have separated out the last six weeks of the calendar on a larger scale.

JAN	about 4500 Million years ago	Formation of the Earth
FEB	4120	
MAR	3775	Oldest known rocks Life exists only as simple organisms like bacteria Blue-greens appear in sea
APR	3400	
MAY	3020	
JUNE	2640	
JULY	2270	
AUG	1990	Rocks in bottom of Grand Canyon formed
SEPT	1510	Protistans appear in sea
OCT	1140	
NOV	750	Sponges appear Jellyfish and marine worms appear Life invades land
DEC	385	First backboned fish Backboned animals move onto land Reptiles rule Rocks at top of Grand Canyon formed Mammals appear

Date	Million years ago	Event
16 NOVEMBER	557 Million years ago	
17		
18	533	
19		
20	508	
21		
22	483	
23		
24	459	Simple land plants begin to grow
25		
26	434	
27		
28	410	First backboned fish
29		
30	385	First animals move to land — millipedes
1 DECEMBER		Fish begin to leave water
2	360	
3		Insects and amphibians develop on land
4	335	
5		
6	310	
7		First reptiles appear
8	286	
9		Trees begin to bear cones
10	262	
11		
12	238	
13		First mammals appear
14	210	
15		
16	185	
17		Reptiles rule the land
18	160	
19		First birds take to the air
20	136	First flowering plants begin to bloom
21		
22	112	
23		
24	86	
25		
26	62	Dinosaurs disappear
27		
28	37	Mammals greatly increase in numbers
29		
30	12	
31		Modern man appears

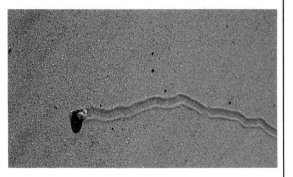

we are told, the Earth was formed, the Solar System and Universe began, life originated and dinosaurs lived where we live now.

It is relatively easy to appreciate the way in which a historian or an archaeologist dates his study objects, whether it be by reconstructing an ancient American Indian encampment, for example, or reading the ancient records of Babylonia. It is not too difficult for laymen to follow such lines of evidence. But the methods of the geologist working out the history of the Earth are a mystery to many people, and so often there is a certain scepticism about his reconstructions. The history of the Earth and the formation of rocks are so far back in time that the geologist is constantly asked, 'How can you be sure?'

The geologist's basic unit in dealing with the Earth's history is a million years, and the astronomer's unit for the Universe is a light-year (186,000 miles per second). To appreciate these vast time-spans for measuring the Earth's history and to put them into relative dimensions, perhaps we need the help of some illustrations.

Imagine the history of Planet Earth as represented by a single calendar-year. The Earth is about 4,500 million years

On the left is a trail made by a worm-like animal about 550 million years ago, fossilized in the upper surface of a sandstone bed in Peary Land, Greenland. On the right is a trail made by a similar animal, a gastropod, in the sand of a modern beach.

This front half of a fossil fish was found in a core sample drilled about 12,000 feet below the North Sea, in the South Brae oil field off Britain. It is about 150 million years old, from the Jurassic Age.

old, so each day of the year is equivalent to about 12.3 million years. Some of the major events that have taken place during Earth's history are marked on the annual calendar. Representation in this way clearly illustrates the vast time-intervals between the various events. We can see that modern man appeared on Planet Earth only about 35,000 years ago, which on the annual calendar representation means he arrived just in time to celebrate 'old year's night' on 31 December.

Try thinking instead about a visit to one of the Earth's most dramatic and inspiring natural wonders, the Grand Canyon of the Colorado River in Arizona, USA. This mile-deep hole in the Earth's surface can be used to illustrate the time-scale of geological events. The Grand Canyon is deeply incised into a thick geological sequence of little-deformed Proterozoic sediments, and it has long been considered a promising locale for the detection of fossil evidence of early life.

Imagine walking down the Canyon on the mule tracks (no doubt the temperature will be at least in the mid-nineties Fahrenheit by midday!). The rocks around the rim of the Canyon are about 200 million years old, and will not contain any fossil records of mammals or birds, but only traces of reptile remains and often imprint-fossils of fern leaves and insect remains. Moving down the Canyon, at about halfway

THE GEOLOGICAL COLUMN

Geological history is normally represented by time-scales of the Geological Column. For stratigraphical purposes the scales are based on data drawn from the study of fossils, plant pollen and spores, and also from geophysics and geochemistry. The age of the most ancient rocks is determined by radiometric methods (see *Physical Clocks*), and chronological scales are used for putting together the sequence of events in time. The periods of the PreCambrian are based entirely on radiometric dating; they include both metamorphic complexes and sedimentary rocks.

The more recent periods of the Neogene and Quaternary are quite complex and are not included in the Geological Column. There are currently fifteen stages of the Quaternary and eight of the Neogene.

Different formations around the Earth often give slightly different periods and systems for the rocks within the same era, so geologists working in different areas may not fully accept this Geological Column, which is based on names most commonly used in North-west Europe. As new geological data is accumulated, the Geological column is slowly being adjusted or confirmed.

This artist's impression gives an idea of the ages of the different layers of rock in which fossils and microfossils are found. Fossils are preserved in ancient sediments throughout the Geological Column. Fossil plant spores are very resistant to being broken down chemically or biologically. The study of pollen and spores (palynology) is a basis for dating the different geological strata. Both these examples are from the Upper Carboniferous Age, about 280 million years ago. Lagenostoma (top) is from a British coal ball. Lagenicula horrida **was found at Grand Ledge, Michigan.**

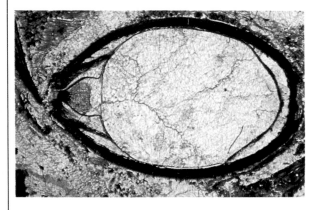

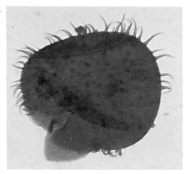

THE PRECAMBRIAN PERIOD

There is an alternative way of naming the geological intervals to the one used on the chart. It was suggested by Harrison and Peterman in 1980.

PRECAMBRIAN

Archean   4,500–2,500 million years ago
Hadean	4,500–3,900
Early Archean	3,900–2,900
Late Archean	2,900–2,500

Proterozoic   2,500–570 million years ago
Early Proterozoic	2,500–1,600
Middle Proterozoic	1,600–900
Late Proterozoic	900–570

CAMBRIAN AND YOUNGER

570 million years ago to the present

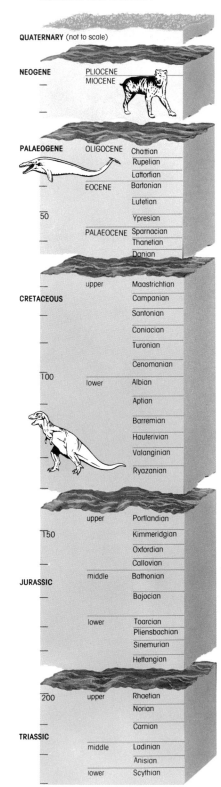

MESOZOIC AND CENOZOIC

QUATERNARY (not to scale)

NEOGENE — PLIOCENE / MIOCENE

PALAEOGENE — OLIGOCENE: Chattian / Rupelian / Lattorfian
EOCENE: Bartonian / Lutetian
50 — Ypresian
PALAEOCENE: Sparnacian / Thanetian / Danian

CRETACEOUS
upper: Maastrichtian / Campanian / Santonian / Coniacian / Turonian / Cenomanian
100 —
lower: Albian / Aptian / Barremian / Hauterivian / Valanginian / Ryazanian

JURASSIC
150 —
upper: Portlandian / Kimmeridgian / Oxfordian / Callovian
middle: Bathonian / Bajocian
lower: Toarcian / Pliensbachian / Sinemurian / Hettangian

TRIASSIC
200 —
upper: Rhaetian / Norian / Carnian
middle: Ladinian / Anisian
lower: Scythian

PALAEOZOIC

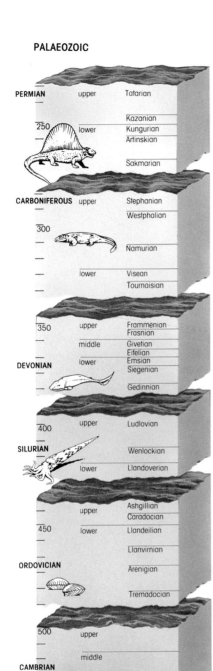

PERMIAN	upper	Tatarian
		Kazanian
250	lower	Kungurian
		Artinskian
		Sakmarian
CARBONIFEROUS	upper	Stephanian
		Westphalian
300		Namurian
	lower	Visean
		Tournaisian
350	upper	Frammenian
		Frasnian
	middle	Givetian
		Eifelian
DEVONIAN	lower	Emsian
		Siegenian
		Gedinnian
400	upper	Ludlovian
SILURIAN		Wenlockian
	lower	Llandoverian
	upper	Ashgillian
		Caradocian
450	lower	Llandeilian
		Llanvirnian
ORDOVICIAN		Arenigian
		Tremadocian
500	upper	
	middle	
CAMBRIAN		
550	lower	

PRECAMBRIAN

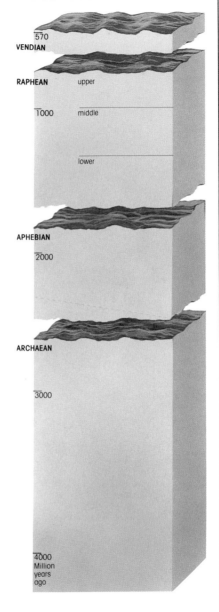

570	
VENDIAN	
RAPHEAN	upper
1000	middle
	lower
APHEBIAN	
2000	
ARCHAEAN	
3000	
4000 Million years ago	

the rocks have been dated at around 400 million years old
and any fossils are likely to be fish remains. Moving farther
down the track (I wish I'd taken the mule!), the rocks become
progressively older and those dated 500 million years old
show no evidence for backboned animals of any kind; the
only evidence for life is shown by shells and aquatic worms
that probably left their remains on the shallow, muddy sea-
floor that existed those 500 million years ago. About three-
quarters of the way down, fossil evidence in the rocks
becomes much rarer and only microflora have been identified. Although the microflora are tiny micro-organism
fossils, they contain diverse microfossil assemblages and at
least eight morphologically distinct types of micro-organism
have been described.

When you finally reach the Colorado River at the bottom
of the Canyon, you will have travelled one mile vertically, but
2,000 million years in terms of the Earth's history. The
ancient PreCambrian rocks at the base of the Canyon contain
trace-fossil remains of micro-organisms which probably
existed as blue-green algae and bacteria in ancient 'oceans'
about 2,000 million years ago.

Time to climb back to the rim. This hypothetical visit will
have illustrated the vast time-span of the last 2,000 million
years' history. But we must also remember that the Earth is
another 2,500 million years older and that none of this early
period of the Earth's history is represented in the Grand
Canyon rock sequences. The bottom of the Canyon represents around mid-June on our annual-calendar scale.

RELATIVE TIME

Most rocks, particularly sediments, cannot be dated with
radioactive isotopes. And even those that can be radiometrically dated often are not, because of the effort and
expense involved. Consequently, geology is dependent on
relative time-scales (time relative to the depositing of
particular types of rock), although these can be integrated
with the radiometric time-scale in some places.

The development of a scale of relative time began in the
eighteenth century. It took some while for the concept to gain
acceptance. At the beginning, Western-European scientists,
who were the leading geologists of their day, were strongly
influenced by the biblical account in Genesis, and in
particular the interpretation that the Earth was only a few
thousand years old and all living plants and animals were
created at the same time. Gradually, however, the scientific

This fossil in a rock is of an early seed fern plant, Alethopteris serli. **It is from the Upper Carboniferous Age, 270–330 million years ago.**

These tiny fossils represent three geological ages. Above right is Triadispora Staplini, **from the mid-Triassic about 220 million years ago; above left is a Fossil Cuticle** (Sawdonia) **showing evidence of the earliest stoma in fossil plants (Lower Carboniferous: 300–350 million years ago); below is a fossil spore,** Triangulatisporites rootsii, **from the Upper Devonian about 350 million years ago.**

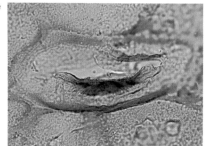

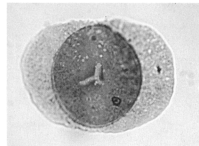

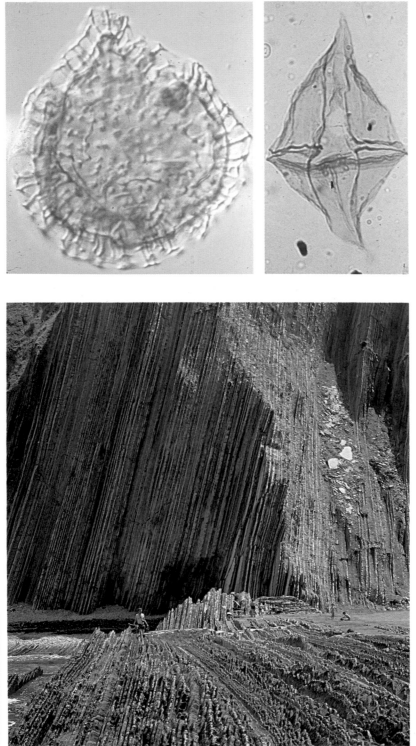

Fossils of Dinoflagellate Cysts, planktonic algae which lived in the open seas about 80–120 million years ago in the Australian Mesozoic.

This dramatic example of rock deformation is in a cliff near Zamaya in Spain. A steeply-dipped interbedding of sandstones, it was probably deposited in muddy waters about 50 million years ago.

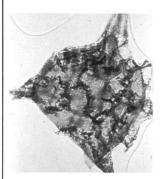

climate of thought, although not rejecting biblical truth, began to integrate geological findings into a fuller and more accurate picture of the age and history of the Earth.

Geologists are able to study rocks exposed at the surface of the Earth or those within reach of a drill. They can obtain information from many different rocks, and these are the source of most of our knowledge of Earth's history. Rocks are divided into three classes according to their origin:

● **Sedimentary rocks**, such as sandstones, were formed by the weathering of existing rocks. The sediments are then carried by water, wind or ice and deposited on land or under water.

● **Igneous rocks**, such as basalt, were crystallized from molten material, either on the surface or below ground.

● **Metamorphic rocks**, such as slate and marble, were formed from the recrystallization of any rock by heat and/or pressure, often deep underground.

Ancient igneous and metamorphic rocks, particularly in the PreCambrian age, have provided useful information on the origin and early history of the Earth's crust. Younger rocks, often deformed by collisions on the Earth's crust and by pressure from igneous bodies, can be used to understand and explain the making of mountain chains, and the overall structure of the Earth's crust, plate tectonics and continental drift. Sedimentary rocks are of particular interest to geologists and palaeontologists (who study fossils), because the rocks and the fossils they contain provide evidence for reconstructing the surface layers of the Earth.

The science of stratigraphy is the study of the sequence of these layered sedimentary rocks found on the Earth's surface. In undisturbed sequences of sedimentary rocks, the youngest sediments will occur at the top of the sequence and the oldest will lie at the base. Sedimentary rocks and sediments cover more than sixty per cent of the Earth's surface. They provide the main record of past environments, climates, movements of coastlines, evolutionary changes and so on. Fossils occur in sediments and they provide much information on the changing patterns of plant and animal life. Palaeontology, palynology (study of plant pollen and spores) and palaeobotany (ancient botany) have provided the accurate basis for and the principal means of establishing what is called the 'Geological Column'.

Measuring Geological Time

Where do the Sun and its planets stand in the time-scale of the Universe? Can events in the Solar System's development be dated and placed in sequence relative to one another? Most importantly, in our inquiries into the origins of life how can we establish a succession of events on Planet Earth and calibrate that succession in years, including the period from the origin of life through to the establishment of mankind on Earth?

These Kangamint Dykes were formed about 2,000 million years ago, when volcanic activity ceased and the molten magma solidified to form vertical dykes. The radiometric age of such dykes is often very different from that of the surrounding rocks.

In the 1650s, a theological scholar, Archbishop James Ussher of Armagh, Ireland, worked out from biblical evidence that the world was created in six twenty-four hour days at 9 a.m. on 26 October, 4004 BC. This precise interpreted date, printed as a margin note in some translations of the Bible, was accepted by many scientists well into the eighteenth century and by the general lay public until the beginning of the nineteenth century. But nineteenth- and early-twentieth-century geologists, often devout Christian men, were unable to accept so short an allocation of time for the Earth's history.

Looking around, these geologists saw a regular, extensive and predictable succession of fossils and rocks throughout the Earth. They discovered that succession to be consistent with other criteria for ordering a sequence of events which happened before, after and at the same time as other events. Vast sequences of stratified sedimentary rocks were observed, with thicknesses of hundreds of thousands of feet. And these contained high concentrations of fossils that changed, generally speaking, from relatively simple organisms of low diversity in the lower, older sections of the rocks to more complex and more highly diverse organisms higher up the geological column. The rates of sedimentation and biological change were carefully measured, and the geologists concluded, even with often poorly understood and incomplete geological succession of strata, that these huge successions of rocks were the product of much greater geological periods than a mere few thousand years.

In recent years enormous progress has been made in studying many aspects of geology, and precise and accurate methods have been developed for measuring the age of geological materials (geochronology). A major step forward in geochronology was the discovery of radioactivity at the turn of the present century; this marked the beginning of a new era in geological studies. These dating methods are based on the radioactive decay of certain nuclides with very long half-lives and are applicable over virtually the entire range of geological time (approximately 4,000 million years). The methods can be used in a wide variety of commonly occurring rocks and their constituent minerals.

HOW ACCURATE?

Such dating methods applied to ancient rocks give ages of millions and thousands of millions of years. Such large ages can be somewhat bewildering to readers who are not acquainted with the techniques and results. Absolute

WHAT IS RADIOACTIVITY?

Many atoms that occur in nature have unstable nuclei which decay spontaneously to a lower energy state. The atoms are present in radioactive elements and the decay is called radioactivity. When a radioactive atom decays, it changes to another kind of more stable atom.

There are several ways this decay happens:

● In **alpha-decay**, the nucleus of the parent atom loses two protons and two neutrons, the mass number of the element is reduced by four and the atomic number decreases by two.

● In **beta-decay**, the nucleus emits a high-speed electron, one of the neutrons changes into a proton and the atomic number increases by one.

● In **electron capture,** the nucleus captures an electron which combines with a proton to form a neutron, and the atomic number decreases by one.

Alpha-decay and electron capture leave the mass number unchanged.

Stabilization is not always a simple, one-step process. For example, Uranium-238 goes through many changes before the stable form Lead-206 is reached.

The number after the element name refers to the number of particles in the nucleus; forms of the same element having nuclei with different numbers of particles are known as **isotopes** of the element, hence the term **isotopic dating**.

Rocks can be eroded,
as they are worked on by wind
and water, to leave spectacular
skeletons of what they once
were. The double arch is from
the National Park in Utah, United
States. The other two are Cedar
Breaks, also from Utah, and
Mount Eisenhower in Canada.

radiometric dating in geology is not just a kind of smart-alec way of producing older and still older ages. It is not even guesswork any more. Such absolute dating is based not only on a number of sound physical laws, but on several parallel methods, by which the results can be checked independently.

Yet this still does not mean these methods are at all easy to apply in geochronology. They require the utmost of modern-day, sophisticated instrumentation and analysis. Also, the rocks of the Earth that contain the radiometric physical clocks have not necessarily always existed under steady, uniform conditions. They have been subjected to many different kinds of processes and changes during the long years of their geological history. It is not nowadays a problem to measure the 'age' of a rock, but sometimes uncertainty arises on the interpretation of what the 'age' means. It may

Natural caste of tree trunk (Giant Club Moss) found at Swillingdon Quarry, UK, and dated as Carboniferous (270–350 million years old).

PHYSICAL CLOCKS

We have mentioned radiometric dating methods for rocks in our main text. How is this done?

All heavier natural elements are unstable, and so are several isotopes of the lighter elements. They are present in rocks today because they decay exceedingly slowly. The age of the rocks in which they are found, and even the age of the Earth, is small compared to their time of decay. They represent a left-over from the amount originally present at the formation of the Earth (or of meteorites).

The decay rate of a radioactive isotope can be expressed conveniently in terms of its **half-life**. It is the opposite quality to the exponential growth of populations and other natural growth curves, where rapid rates of increase are doubled with time. Instead of doubling, radioactive decay halves. Let us build an 'egg-timer' to

measure geological time, and assume that it takes up to 40 million years to measure the event. Suppose you design the egg-timer so that only one grain of sand per day passes from the upper to the lower glass bulb. The original capacity of the upper glass would need to be about 15 billion grains of sand. If the timer is designed so that half the sand falls through to the lower bulb in the first 713 million years, half of the remaining sand (a quarter of the original) in another 713 million years, half the remainder (an eighth of the original) in a third period of 713 million years and so on, at the end of twenty 'half-lives' more than 14 billion measurable years would have elapsed.

This gives some idea what happens as radioactive isotopes decay. For example, when Uranium-235 decays to Lead-207, the amount of lead formed from uranium-decay in

a given time depends on the amount of uranium left. Each radioactive isotope decays at a different rate, so that the half-life by which this exponential decay process is measured must be independently assessed for each radioactive isotope.

Once the half-life of an isotope is known, it is possible to date a rock by discovering how far that particular substance has decayed.

Radiometric techniques of absolute age-determination of rocks have developed to the point where it is often possible to date the same rocks by different techniques. Rocks containing only minor quantities of the radioactive elements and their decay-products can often be dated by both the rubidium-strontium and the potassium-argon methods, while it is often also possible to use thorium- and uranium-lead decay series.

The logs of this petrified forest in Arizona once stood as tall trees in a dense forest, growing about 200 million years ago. Covered by hundreds of feet of later sediment, the trees were preserved under special conditions and dissolved silica was absorbed into the trunks. The silica displaced the wood and formed these petrified structures.

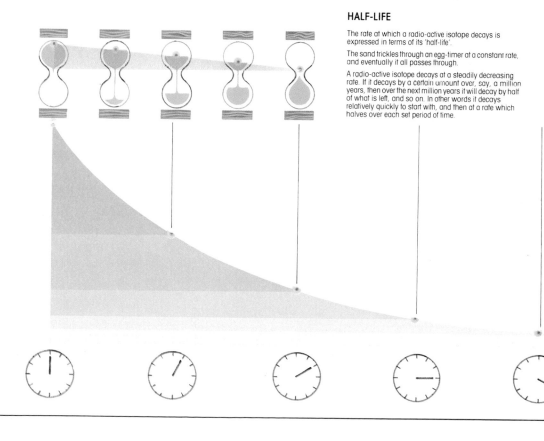

HALF-LIFE

The rate at which a radio-active isotope decays is expressed in terms of its 'half-life'.

The sand trickles through an egg-timer at a constant rate, and eventually it all passes through.

A radio-active isotope decays at a steadily decreasing rate. If it decays by a certain amount over, say, a million years, then over the next million years it will decay by half of what is left, and so on. In other words it decays relatively quickly to start with, and then at a rate which halves over each set period of time.

CARBON-14 DATING

Radioactive Carbon-14 can be used to date events over the relatively recent period of the last 40,000 years — too short a time-interval for correlation using fossils.

Naturally-occurring carbon in the atmosphere and in living plants and animals consists of two stable isotopes, Carbon-12 and Carbon-13 with a natural abundance of 98.892% and 1.108% respectively, while a third, radioactive isotope Carbon-14 is formed by the reaction of cosmic-ray particles with Nitrogen-14 in the upper atmosphere.

The abundance of Carbon-14 in modern wood is 0.000000000107% and it has a half-life (see *Physical Clocks*) of 5,370 years. This is so low that it cannot generally be measured in organic material older than about 40,000 years. Because of its short half-life no existing Carbon-14 isotope can be primordial. Instead it is continually being created in the upper atmosphere about ten miles above the surface of the Earth as a by-product of cosmic-ray bombardment. This continuous flow of newly-created Carbon-14 is quickly incorporated into carbon dioxide, and this enters into Earth's biological cycle via photosynthesis. It is then distributed uniformly throughout the various carbon phases. When Carbon-14 is taken up into biological material, the proportion steadily decreases as the radioactive carbon decays to stable Nitrogen-14. The age can be calculated if the very weak emission of Beta-particles is measured.

Unlike the other radiometric dating methods, radiocarbon dating involves the measurement of radioactivity itself and not end-products. However, in determining ages it is necessary to correct all carbon dates, partly because the rate of formation of Carbon-14 in the past has varied, and partly because of recent additions of 'dead carbon' into the atmosphere from the burning of fossil fuels, such as coal, peat and oil.

Radiocarbon dating can only be used for the last brief portion of geological time. But it has become a most important radiometric method, since a great deal of geological activity has occurred within this short time-span. Wood, peat, shell, bone, hide and fabric may all be dated in this way. Radiocarbon methods have been extensively applied to archaeology and anthropology. Carbon-14 dating helped to disprove the famous Piltdown skull-and-jaw hoax, by showing that the skull gave an age of 620 plus or minus 100 years, while the jaw was only 500 plus or minus 100 years old.

More recent applications of Carbon-14 dating include the rate of sedimentation of ocean sediments, late Pleistocene geology and climate, and also the length of time that meteorites have been on the Earth. Other major geological events, such as the retreat of the last continental ice sheets, accompanying climatic changes, changes in ocean circulation, the post-glacial rise in sea-level and the rise in human civilization, have all been studied using radiocarbon dating techniques.

truly represent the date of original formation. But it could also correspond to some later event in the history of the rock, such as recrystallization or some catastrophe, by which the original data in the clock have been destroyed. The results would then relate to the date of this later event and not the real age of the rock.

It is around these interpretations of geological history and processes that the most serious discussions in radiometric dating now take place. However, we are still safe in concluding that absolute radiometric dating of rocks can be trusted as an approximation of the real age of the rocks, and such techniques are used extensively and successfully in dating ancient PreCambrian rocks on the Earth.

What the Universe is Made of

All studies of the various parts of the known Universe have combined to suggest that the basic constituent elements and elementary particles are the same throughout the Universe. The Universe has a fairly well-defined shape and structure.

Essentially it appears to consist of a very large number of bodies termed galaxies, separated from each other by quite vast distances. What is in these enormous spaces between them? Apparently little or nothing except a high vacuum which contains minute amounts of atoms and molecules such as perhaps hydrogen and dust.

Each galaxy consists of large numbers of stars of varying types, of which our own Sun is apparently an average sort of sample. Presumably many stars have planetary systems like our own Solar System and these in turn may have sub-planetary satellite (moon) systems. These satellites may include not only the large, obvious bodies but also numerous lumps of rock (asteroids), ranging in diameter from millimetres to miles. The galaxies also contain gas-dust clouds which sometimes form recognizable units (nebulae) and sometimes are more diffuse.

HOW OLD IS THE UNIVERSE?

Modern cosmology appears to have shattered Archbishop Ussher's time-scale once and for all. The age of the Universe is estimated at between 10,000 and 25,000 million years old.

The Universe as a whole is continually changing, and in particular it is expanding. Our knowledge of the expansion of the Universe is entirely due to astronomers being able to measure the motion and speed of its luminous bodies. The

Overleaf
All galaxies contain, along with stars and asteroids, great clouds of gas and dust. Some may be stars about to form; others are stars that have exploded. Cygnus Veil is a gas-cloud nebula resulting from a supernova explosion.

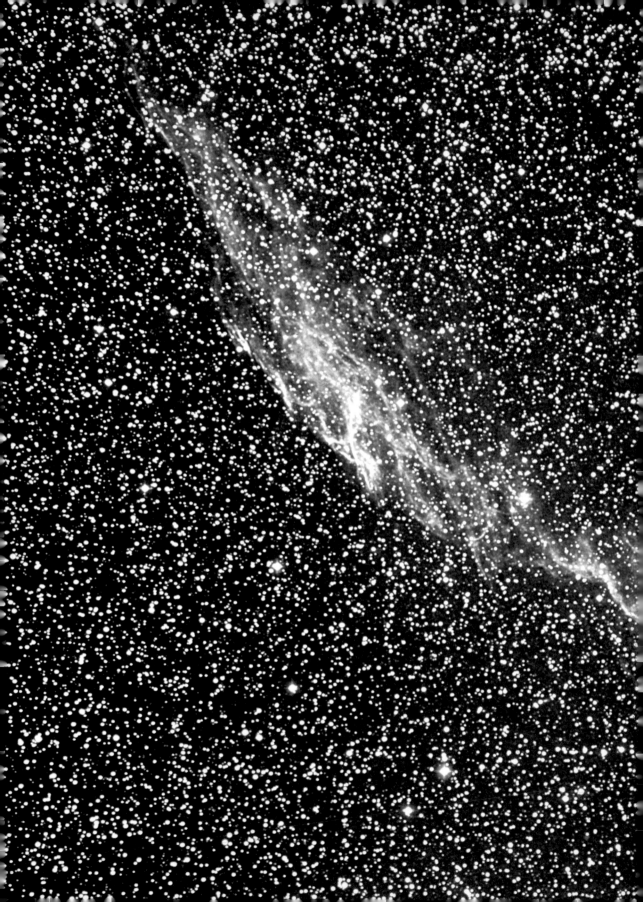

technique used to measure this motion and speed makes use of the familiar property of any kind of wave motion, known as the Doppler Effect.

All the galaxies appear to be moving away from each other at great speeds which, with increasing distance, approach the speed of light. Evidence for this comes, as we shall see, from the red-shift in the spectrum of light emitted by the galaxies. By extrapolating backwards in time it is possible to arrive at an age of the Universe — always assuming such an extrapolation is possible and also that the whole Universe had a specific beginning in time, at the Big Bang or whatever.

In 1923 researches by Professor Edwin Hubble, using the Mount Palomar telescope, led him to formulate the so-called 'Hubble's Law': the recessional velocity of a stellar system is proportional to its distance. When we plot on a graph the recessional velocities of known galaxies against their distances from the Earth, we get a straight-line relationship which is expressed as 'Hubble's Law': $\mathbf{v} = \mathbf{HD}$. ($\mathbf{v}$ is the recessional velocity, $\mathbf{d}$ is the distance and $\mathbf{H}$ is 'Hubble's constant' where $1/H = 10^{10}$ years.) If this straight-line relationship really does hold good everywhere, the strong suggestion is that all galaxies were at a single point some 10 billion years ago.

There are of course errors in the red-shift measurements, and such variations point towards a starting-point for the Universe somewhere between 6,000 and 13,000 million years ago.

A lot of cosmological theory rests on these interpretations of red-shifts and on Hubble's Law. With the passage of time, the 'law' and its consequences have become increasingly accepted. But there are still a few nagging problems, most important of which perhaps is the existence of some galaxies which are apparently as much as 25 billion years old.

Present-day discussion about the value of the Hubble constant revolves around the time-scale of the Universe; it is an argument about whether the beginning of the expansion was 25, 20 or 10 thousand million years ago.

Recent research on the motion of our own galaxy has changed the estimates of Hubble's constant (see S. Weinberg's *The First Three Minutes*). Many of the galaxies on which these estimates are based are in the Virgo cluster, so that if our galaxy is actually travelling towards this cluster, then the velocity of its galaxies due to the general expansion of the Universe must be rather larger than would be inferred

THE DOPPLER EFFECT

You can easily observe the Doppler Effect — in terms of sound waves — if you stand near to a motorway, railway or airport. Notice that the engine of a fast machine sounds higher-pitched (that is, of shorter wavelength) when the object is approaching you, than when it is moving away.

In the diagram, **A** shows the helicopter standing still; sound waves from the engine reach the observer at their normal frequency. In **B**, the helicopter approaches the observer moving a distance **x** between two successive waves; to the observer, the wavelength seems shorter and the frequency higher. In **C**, the helicopter moves away from the observer, now moving a distance **y** between successive sound waves; to the observer, the wavelength seems longer and the frequency lower. The helicopter's speed is greater in **C**, hence the distance **y** is

greater than **x**, and the wavelength and frequency changes are correspondingly greater.

The physical principle applies equally to sound and light waves. The light emitted by star **D** appears bluer (**E**) or redder (**F**), depending on whether it moves towards or away from the Earth.

When an astronomer observes a light wave from a source at rest (**D**), the time between the arrival of wave-peaks to his instrument is the same as the time-interval between the peaks as they leave the source. But if the source is travelling away from the observation instrument (**F**), the time between arrivals of successive wave-peaks is greater than the time between their departures from the source, because each peak has a little farther to travel on its journey than each previous wave-peak. The source moving

away from the observer will appear to have a longer wavelength than if the source had remained stationary. Similarly, if a source is travelling towards the observer (**E**), it will appear to have a shorter wavelength.

This fundamental law of physics may be a little difficult to understand, but a useful pictorial illustration by Steven Weinberg, in his book *The First Three Minutes*, gives an appreciation of the Doppler Effect. It is as if a travelling salesman were to send a letter home regularly once a week during his travels. While he is travelling away from home, each successive letter will have a little farther to go than the one before, so his letters will arrive a little more than a week apart. On the homeward leg of his journey, each successive letter will have a shorter distance to travel so they will arrive more frequently than once a week.

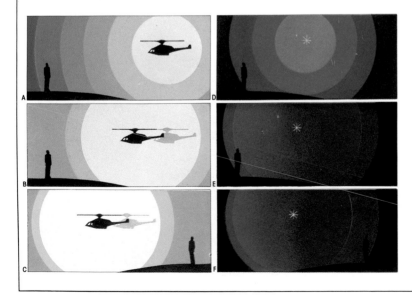

from observing their red-shifts. Weinberg concludes that the Hubble constant is somewhat larger than has previously been estimated — say 20 miles per second instead of 10 miles per second per million light years. However, this conclusion is not universally accepted.

Where do these variations and disagreements leave us? Sir Bernard Lovell has reminded us that these uncertainties of a factor of two (or possibly two and a half) in the age of the Universe do not obscure the fundamental evidence that the Universe is expanding. Sir Bernard goes on to say: 'There would seem to be an inescapable logical conclusion from this evidence that the beginning of the expansion was of a Universe totally different from that which we view today — that it must have been localized, and of immense density.'

CREATION?

Although there are still a large number of scientific arguments to be settled, today the evidence for the Big Bang is considered overwhelming. Modern cosmology has extensively studied microwave radiation in the outermost parts of the known Universe and the red-shifts reveal what we would expect on this hypothesis — distant galaxies are expanding away from our galaxy at immense speeds approaching the speed of light.

Science has pointed very clearly in the direction of the Big-Bang creation of the Universe, but the cause of the Big Bang still remains very much an enigma.

Through history, many theologians have sought philosophical reasons for believing in God. They have often concluded that there must be a 'first uncaused cause', and this they have boldly identified with God. The Big-Bang theory of the beginning of the Universe seems at first sight to satisfy their 'proof'. Scientific explanations come to an end, providing a God-given opportunity to ascribe the Universe to a transcendent cause. This can amount to saying that science had uncovered the secrets of the Universe and disclosed the work of God.

Such reasoning is incorrect and unnecessary and is often called the 'God-of-the-gaps' approach — where Christians or others claim room for God only in areas where human knowledge has not reached. This 'God of the gaps' is a wrong and pathetic substitute for the infinite, all-powerful God of the Bible, who is present within every part of his Universe and yet at the same time immeasurably transcends our every thought. To find God only where there are gaps in our scientific knowledge is unnecessary, misleading and indeed blasphemous. Since the gaps are rapidly decreasing, it suggests a shrinking God. But the God of the Bible is not limited. Either he is there in the whole Universe and in every part of it, or he is not there at all.

We must in fact be very cautious about using cosmology to support the Christian view on the beginning of the Universe. There are several reasons for saying this:

● **A cosmologist can say nothing about transcendent causes.** By definition these lie outside scientific analysis and proof.

● **The argument from an uncaused cause is never totally dependent on there being a 'beginning'.**

● **The Big Bang can be argued either way.** Scientifically it can be convincingly argued — just as convincingly as the reverse — that the Universe originated out of nothing of its own accord, as a result purely of physical laws. In this way, a rationalist's view of creation is equally as acceptable intellectually as its Christian alternative.

Although cosmological theories remain inconclusive, scientific method points clearly towards a Big-Bang origin of the Universe. Science supports the picture of an expanding Universe and reveals something of its beauty, vastness, mystery and antiquity. Christians can equally contemplate the Big Bang and its aftermath and see there the planned work of God; there is nothing at all discordant with science

The arguments about how the Universe came into being can never remain purely mechanistic. We are bound to relate the discussion to ourselves living on Planet Earth. How did the development of the Universe reach the point where life began on Earth?

THE HIDDEN UNIVERSE

The Universe may prove to be a hundred times more massive than can be seen in bright stars and galaxies. The latest cosmological theories are attempting to predict the form of the invisible Universe.

For the last fifty years or so, galaxies have been seen as the building blocks of the Universe and have provided the main clues to an understanding of its ultimate structure. Recent advances in cosmology have progressed to the point where it seems that the bright stars that make up the visible galaxies represent only a small proportion of the matter in the Universe. All astronomers, including conservative ones, are now facing up to the remarkable discovery that even quite nearby regions of space must contain between ten and a hundred times more mass than has previously been accounted for. Professor Michael Disney, in his book *The Hidden Universe*, suggests that around and between galaxies like our Milky Way, space appears to be dominated by massive invisible agencies which defy astronomers' efforts at detection. These appear to contradict present theories of the Universe.

These invisible agencies have been suggested as the only possible means whereby galaxies can be held together. Spiral galaxies of the sort containing Planet Earth and our Solar System would fragment into lumps, and the clusters into which most galaxies are huddled would rapidly disperse into open space, if they were not all held together by the gravitational effect of overwhelming amounts of invisible mass.

Most cosmologists are confident that the missing mass cannot be in the form of ordinary matter. Cosmologists have been able to explain convincingly the chemical composition of the Universe and in particular the relative proportions of the predominating light elements such as hydrogen, helium and deuterium (an isotope of hydrogen). These are natural consequences of thermonuclear reactions that should have taken place during the first three minutes after the moment of creation, when the Universe was still in a super-dense state. The rules of nucleosynthesis are sufficiently well-understood for cosmologists to be able to calculate with confidence the proportion of 'ordinary' matter that could have been synthesized in this way. The permissible range is between five and ten per cent of critical density. Professor J. Silk (University of California) suggests that all this matter could be in the form of ordinary stars without embarrassing the observers too much, although it is quite likely that some exists in the form of hot gas that has not been condensed into galaxies, or as stars which have low mass and are too faint to be seen. But that still leaves ninety per cent of the critical mass of the Universe unaccounted for.

These observations no longer leave a safe position to which cosmologists can retreat behind their theories. We now find ourselves in a Universe largely consisting of invisible mass whose nature can only be wondered at. The search will continue.

The exact identity of the ninety-nine per cent of the Universe that cannot be seen is

still unknown. But certain exotic elementary particles (gauge particles, for instance) are being suggested by particle physicists and cosmologists. The details of this new picture of the formation of galaxies have yet to be fully explored.

Such studies may yet confirm the reality of some of those 'theoretical' elementary particles that physicists speculate about.

The Milky Way, our own galaxy, contains some 100,000 million stars with an enormous range of mass, temperature and brightness. It is now thought that around and between such galaxies there may be a 'hidden Universe'.

in that. It will never be possible, or even necessary, to prove scientifically from cosmology that it *must* be the work of God. But certainly the Big-Bang theory fits well with the idea.

The Bible asserts that the Universe was created, or called into being, by God. It does not exist in its own right. Everything, according to Scripture, owes its existence to God.

This puts a very interesting question before us: When did history begin? Dr Francis Schaeffer remarks that if we are thinking with the modern concept of a space-time continuum, then obviously time and history did not exist 'in the beginning'. But if we think of history in contrast to a static eternal state, then history began before the creation as recorded in Genesis. When we think back before 'In the beginning God created the heavens and the earth', we are not left with something floating around in a vacuum. Something existed before creation, and that something was not static and impersonal but a personal God, a Father who loves us. There was a plan, there were communication and promises before the creation of the Heavens and the Earth. And there was Christ, the key to everything, who existed 'before the world began'.

Everything, at each of the levels of the origin and development of the Universe, fulfils the purpose of creation. The mechanical, scientific part of the Universe acts in perfect harmony. For the Christian the fulfilment of God's plan and purpose in the creation of the Universe can be clearly seen in four distinct areas:

● **The Universe has order;** it is not a chaos. Scientists are able to proceed from the particulars of being towards an understanding of the unity of being.

● **The revealed nature of creation speaks of existence itself.** The Universe is there; existence is there; God is there.

● **The Universe speaks of God's character.** God not only exists and is a God of order and reason, but God is good. He created a Universe that is basically good, even if the Earth is spoilt in some aspects.

As the Creator, God shapes and fashions and brings into existence what he has planned. And when God is finished with this process, what he has made speaks of God the Creator.

The Stars and their Origins

Stars can be broadly classified into two types: Population I and Population II types. The stars at the centre of the galaxies are known as Population II stars. They are produced from primeval hydrogen and consequently are relatively poor in metals. Population I stars occur at the outer edges and in the spiral arms of the galaxies. They have been formed from dust ejected from Population II stars after explosion (Nova and Supernova), together with a hydrogen–helium matrix, and consequently are relatively rich in higher elements including metals. Our own Sun is such a Population I star.

In addition to stars, the galaxies also contain large quantities of gas and dust, and this material is constantly being converted into Population I stars. Evidence for quite recent formation of stars (1.5 million years ago) has come from observations of certain star associations (such as Zeta Persei in Perseus), which must have had a common centre some 1.5 million years ago. Another example occurs in Orion, with a common centre 2.8 million years ago.

The Population I star is considered, then, to form by condensing of dust and gas drawn together by gravitational attraction; this gives initially relatively dark globules.

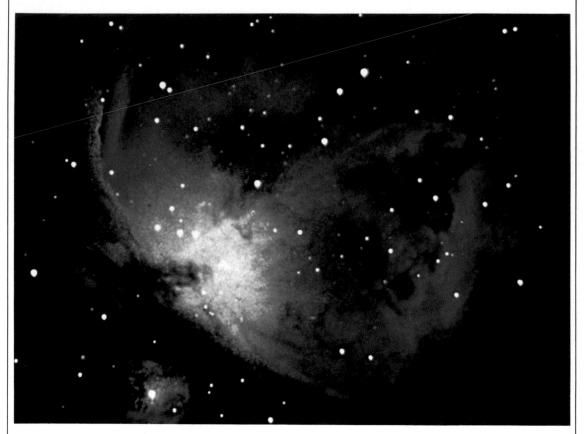

Astronomers can observe such 'Bok's globules', especially when viewing them against a luminous background as in the Rosette nebula. And there is evidence to show that stars are constantly being born by processes of this type. This has come from examining other celestial bodies, including the so-called Herbig-Har objects — for example the T. Tauri star observed in a dark cloud near the Orion nebula. These so-called T-association stars have peculiar spectra and contain large amounts of lithium. They appear to be stars in the making, probably undergoing gravitational contraction.

A plot of star magnitude and spectral class of many stars shows that stars fall into specific classes. These include:

● The recently discovered **quasars** (enormous energy-emitters) densely packed (several tons per cubic inch). They are very distant, and may represent matter as it was early in the life of the Universe.

● Then we have a broad band of the **main sequence stars**, which include our own Sun.

● These in turn, as they grow older, cool down and

The Orion Nebula is a region of active star formation.

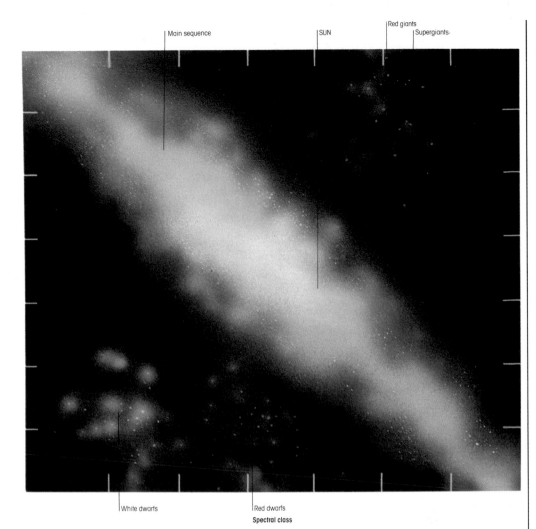

Main sequence　　　　　SUN　　Red giants　Supergiants

Absolute magnitude

White dwarfs　　　　Red dwarfs

Spectral class

A plot of the magnitude and spectral class of many stars shows that stars fall into specific classes. Our Sun is a main sequence star.

enlarge. They begin to radiate a redder light and become **red giant stars**.

● These ultimately explode in a **supernova**, scattering their material into space to provide new fodder for the production of new Population I stars. But they leave behind a small core which becomes the **white dwarf** or neutron star, produced by some implosion process.

● These finally give rise to **dark dead stars**.

HOW GALAXIES AND STARS WERE FORMED

Galaxies are generally presumed to form from a primeval hydrogen gas cloud. It is of great importance for an understanding of the Universe to know why it should consist of essentially similar-sized large bodies separated by high vacuo. Why is it not just a finely-dispersed cloud of gas and dust?

If we assume that the Big-Bang creation produced roughly similar-sized lumps of aggregated matter, then there is no problem.

If, on the other hand, we assume that the tenuous gas extended over light-years, then we have to postulate that particles draw together by mutual gravitational and magnetic attraction to form an enormous but increasingly unstable mass. This in turn begins to break up by processes, say, of condensation and rarefaction into smaller clouds.

This process can continue as more and more sub-clouds are formed from each parent, until a sub-cloud structure is reached in which the kinetic energy makes it rotate, and this rotatory motion will tend to increase as it gets smaller. The type and degree of initial rotation will determine the shape of the galaxy, whether flat disc-shaped or thicker and more ovoid in structure, when gravitational and rotational forces are finally balanced.

As the cloud shrinks the liberated energy will cause the temperature to rise. At lower temperatures (10,000–25,000°C) little or no energy is radiated as heat, provided the cloud is of a size much greater than that of 10,000 million average suns. The cloud cannot therefore shrink further, since if it tried to do that, the energy content would simply expand it to its initial size. Consequently the cloud must fragment into smaller clouds, and when these reach the size of 10,000 million suns (the average galactic mass), half the gravitational energy is released and radiated as heat. The cloud continues to shrink until its size is about 100,000 light-years in diameter, when it becomes too dense to undergo further shrinkage and so fragments again into sub-clouds. These are the embryo Population II stars.

Mechanisms such as this can at best be pale shadows of reality, but they give some idea of the type of possibilities available. The resulting galaxies vary in size from our own (100,000 light-years in diameter) to small ones such as the Andromeda (4,000 light-years across), the average size being about 12,000 light-years. These variations are relatively small ones. The numbers of contained stars, however, appear to vary more. The large galaxies may have up to 200,000 million stars and the dwarf ones only about 300 million, with the average about 10,000 million stars.

The Origin of our Solar System

How and when did Planet Earth and her sister planets begin? What is their relation to the Sun and other stars? Have they always existed in a solid state, with liquid water and a gaseous atmosphere envelope surrounding each one?

Scientists (and others) have asked such questions throughout humanity's time on Earth. Studies down the ages — by scientists, sailors and even shepherds during their night vigils — have come to conclude that the Earth is an integral part of the Solar System.

The origin of our Solar System is intimately connected with the nature of our galaxy, just as the origin of our galaxy resulted from the underlying structure of the whole Universe. This means that all our explanations of the origins of the Earth must be consistent with everything known or predicted, not only about the Earth, Moon and Sun, but also about the eight other planets and their satellites, some many thousands of asteroids, and numerous meteorites, comets and possibly interstellar dust.

At every stage from the Big-Bang creation of the Universe down to the current state and environment of Planet Earth, there appears to be a close interrelationship with previous cosmic events.

Why are we the kind of creatures we are? It is often suggested, with some scientific support, that the answer depends not just on the kind of planet we live on or our position relative to the Sun, but also on the departed physical nature of the first beginnings of the Universe itself. Philosophers tell us that mankind is a creature of the

The NASA Skylab took this photo
of a flaring outburst on the Sun.
To get some idea of the scale of
these flares, remember that the
diameter of the whole Sun is
about 110 times that of the
Earth.

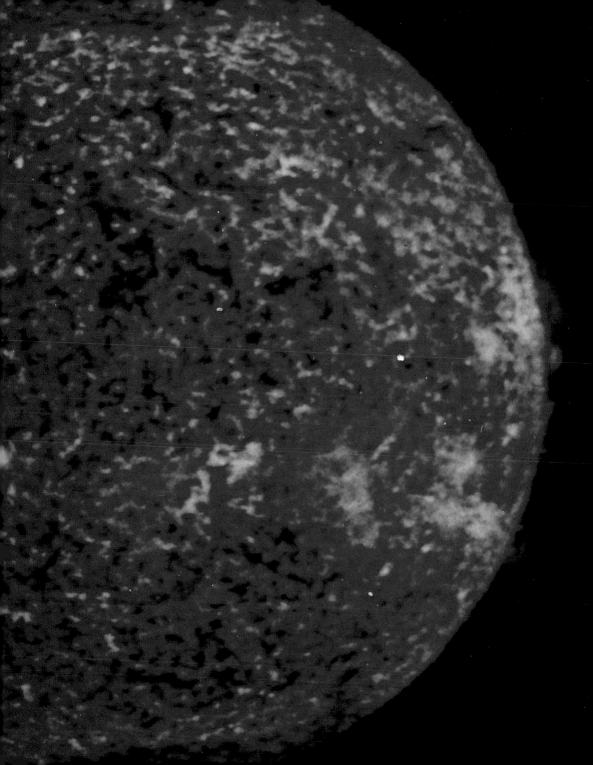

Universe and creatures like us could not have evolved if the Universe were other than it is.

Looking around at our fellow men and women, this somewhat dull, sterile view of mankind, of our origins and purpose on Earth, seems a little impersonal. It is only right and natural that we should ask, 'Was humanity's origin a routine, unplanned event in the natural processes of the origin and development of the Universe? Or is there a purpose and plan for human existence on Planet Earth?'

When we consider the origin of our Solar System, much more specific and detailed issues are involved than general, broad-sweep questions on the origin and nature of the Universe. Those can only, at this time, provide 'easy' answers that do not deal with specifics. Questions about our own Solar System are much more specific, and scientific experiments have been designed to test and measure theories about its origin and the relationship of Planet Earth to the other planets.

Most theories of planetary origin have special reference to our Solar System, and see it as a special case of the almost universal process of binary-star formation. Any theory of the origin of the Solar System has to account for the various specific characteristics of the system:

● **The planets revolve around the Sun in the same direction** in circular orbits which lie almost in a single plane. The planets also revolve around their axes in the same direction as the orbital revolutions. The planetary orbits lie largely in the equatorial plane of the Sun.

● **The distance of the planets from the Sun follows the Titius-Bode 'law':** $(r_n = 0.4 + 0.3(2)^{n-1}$, where r_n is the distance of the nth planet from the Sun using the Earth's distance as unity). The outer planets Neptune and Pluto, however, fail to follow the law.

● **The rotation of the system is concentrated almost wholly in the planets and their satellites**, whereas the mass of the system (99%) is concentrated at the centre in the Sun, which has only 2% of the angular momentum.

● **The four inner planets have low masses but high densities** (about 4000 kg m^{-3}), whereas **the four outer planets have high masses and low densities** (about 1000 kg m^{-3}), with Pluto (density about 4000 kg m^{-3}) an exception. Presumably a corollary to this would be that the inner planets contain a greater proportion of the heavier elements.

● As far as **Planet Earth** is concerned we must explain

the formation of the particular and characteristic abundance of elements and molecules which make up its crust and atmosphere.

HOW DID THE PLANETS FORM?

One of the earliest and best-known theories of planetary formation is the nebular theory. This was put forward by Laplace in 1796, and with various modifications it still remains one of the more attractive hypotheses. The theory suggests that the Solar System originated from a rotating gaseous-dust mass (nebula), which spun off gaseous rings. The material of the rings condensed further into planets, while the central portion remained and after suitable contraction became the Sun.

A major difficulty in this theory concerns the angular momentum of the planets and of the Sun. It appears that the nebula could not in fact have spun sufficiently fast to fling out rings. If we were to add the current measured angular momenta of the planets to that of the Sun, the result would be to increase the Sun's rate of rotation by about fifty times — to about one revolution per half-day, instead of the actual one revolution per twenty-five days. However, this faster rotation would only result in an increased centrifugal force at the Sun's equator of about five per cent of the gravity force, which would be insufficient to overcome the force of gravity and to throw off rings. So our calculation suggests the Sun could not have flung off the planets centrifugally.

These general conclusions led to a temporary abandonment of the Laplace theory (until 1943) and its replacement by alternative theories, which served to explain the enhanced angular momenta of the planets relative to the Sun by invoking a collision or near-collision involving one or more large bodies closely approaching our nebular mass and causing portions to be torn away, to fall ultimately into appropriate circular orbits with momenta enhanced by the process.

However, since 1943 the original objections to the Laplace theory have been largely overcome, and it is now possible to account for the differences in angular momenta of the Sun and planets. There are numerous modern variations on the Laplace mechanism postulated by Alven, Weizsacker, Kuiper, Hoyle and others. They mostly suggest that the primordial Sun had a moderate magnetic field, unlike the very weak field possessed at the moment.

The theories differ in the amount of field strength

THE INNER PLANETS

Mercury
A third Earth's diameter, and three
times nearer the Sun than Earth
Visited by Mariner 10 in 1974

Venus
Slightly smaller in diameter than
Earth, two-thirds of Earth's distance
from the Sun. Visitors: US Mariner 2
and Soviet Venera spacecraft

Earth's moon
Largest satellite in proportion to its
parent planet

Earth
Surface approximately $3/8$ land and
$5/8$ water, atmosphere mixture of
nitrogen, oxygen and other gases
Only solar planet that could possibly
support life

Mars
About half Earth's diameter, and half
as far again from the Sun. Two small
satellites

THE OUTER PLANETS

Jupiter
Separated from the inner planets by
a wide asteroid belt. Its mass is
twice that of the other planets put
together. More than five times further
from the Sun than us. Note its great
red spot

THE SUN

27 000 light years from the centre of
its galaxy. 109 times Earth's
diameter, and 333 times its mass
Its surface temperature is reckoned
to be about 6 000°C

Saturn
Second largest solar planet, almost
twice as far out as Jupiter, like
Jupiter, mainly made of gas

Uranus
Four times Earth's diameter, twice
Saturn's distance from the Sun
Methane atmosphere makes it 'the
green planet'

Neptune
Slightly larger than Uranus, the
farthest out of the giant planets. Also
made of gas. Its larger satellite,
Triton, is the largest in the solar
system

Pluto
Remotest and least-known planet,
only detected visually in 1930.
Nearly as small as Mercury

required. Thus Alven requires an enormous field (about 300,000 gauss) compared with the Sun's probable very weak field of one or two gauss, which leaves the problem of explaining the subsequent loss of field. Hoyle, on the other hand, requires only a small field (one gauss). The point of postulating this magnetic field is that it would have slowed down the central portion of the protosun and increased the speed of the outermost edges, so increasing the centrifugal force.

If the general theory of planetary systems is correct, the angular momentum of a star must be low and that of its planets high. Attempts have been made to compare the spin-rates of stars with that of our Sun. This should make it possible to recognize planet-bearing stars. These studies showed that stars which are massive, hot and quickly-evolving are rotating rapidly, while most main-sequence stars (including our Sun) are rotating slowly. Where has all the angular momentum gone? The suggestion is that it went into planets. It has been calculated that our galaxy contains

This artist's impression tells the story of how Planet Earth developed from a cloud of gas and dust to the rocky planet we live on. The story has taken about 4,600 million years to unfold.

some 10^{11} stars, and by this criterion almost all should possess planetary systems.

None of the above is entirely satisfactory and the theorists are usually handicapped by the lack of sufficient data.

As recently as 1984, Professor Tom Gold has provided astrophysicists with a new agenda and a new theory on how our Solar System was formed. Gold suggests scientists should abandon the conventional framework; the idea will have to go that the Solar System was formed perhaps only in some tens of thousands of years, by rapid condensation within a molecular cloud, and that the Sun and planets were then formed by differential condensation within such a compact mass. Gold proposes that the condensation was much slower, perhaps extending over several hundreds of millions of years. It is early days yet to know whether this theory will be widely accepted.

Even today, with the tremendous advances in astro-physics, many of the theories about the origin of the Solar

System are unsatisfactory. Disagreement in detail among the scientists is still rife; an agreed theory is probably not yet forthcoming.

LIFE ON EARTH ONLY?

It is possible, however, to paint a general picture of what happened thousands of millions of years ago. The galaxy was at least a few thousand million years old when a cloud of dust and gas, buffeted by the explosive birth — or perhaps death — of a nearby massive star, collapsed through interstellar space. Within a spiral arm of the Milky Way, about two-thirds of the way out from its centre, the nucleus of this rolling cloud of dust and gas became dense and hot, ignited and became a star — our Sun. The remaining particles probably gathered into rings around the newborn Sun. About 4,600 million years ago, these particles were drawn together, ring by ring, to make up nine planets, at least thirty-three moons, many thousands of asteroids and millions of meteorites and comets.

Contemporary opinion inclines to accept a nebular theory for the origin of the planetary system, and it is believed that the Sun acquired the nebula as a natural consequence of its condensation from the primeval gas cloud. Sir Bernard Lovell pointed out in 1970 that if this interpretation is correct then a most important consequence is that planetary systems around stars must be a common feature of the Universe — and thus the Solar System can no longer be regarded as effectively unique because of the extreme rarity of stellar encounters. Lovell goes on to point out a host of difficulties and uncertainties that remain, not only in regard to the detailed processes by which planets form from a solar or stellar nebula, but also because the existence of planetary systems around stars other than the Sun is still only theory: there is not yet decisive observational evidence for it.

Nevertheless the circumstantial evidence is considerable, both for the processes of stellar formation from interstellar clouds and for the existence of planets around some of the nearer stars. A Universe in which there exists a multiplicity of planetary systems has become a common feature of modern astronomical thought. There are many who believe also that it is not unreasonable to ask whether living organisms may have developed in planetary systems other than our Solar System.

The Solar Planets and Moons

I t is time now to look in more detail at the Sun, its family of nine planets, their attendant moons and the asteroids. Their formation from that interstellar cloud is thought to have taken place between 4,500 and 5,000 million years ago.

The innermost of the nine planets, **Mercury**, is similar to and scarcely larger than the Earth's moon. The second planet, **Venus**, has a near-circular orbit; its barren, rocky surface is permanently hidden beneath thick clouds of vapour. **Planet Earth**, which is about the same size as Venus, has a single **moon** satellite which is the Solar System's sixth largest. The fourth planet, **Mars**, is bigger than Mercury, but smaller than the Earth. It contains a spherical red desert, has polar ice caps and two tiny moons (Phobus and Deimos). Mars is the outermost member of the group called the 'inner planets'.

Beyond the inner planets are great masses of flying rocks, some over 600 miles in diameter, called the **asteroids**; they are concentrated in a series of bands.

Farther out from these asteroids, and by far the largest of the planets, the fifth from the Sun, is the immense cloud-banded world of **Jupiter**; her diameter is more than eleven times that of the Earth, and she has at least thirteen moons. Slightly smaller than Jupiter and beyond it is the ringed planet, **Saturn**, with ten moons of which the largest, Titan, is bigger than Mercury. The diameters of the second pair of giant planets, **Uranus** and **Neptune**, are both about four times that of the Earth. Uranus has five small moons. Triton, one of Neptune's two moons, is the largest in the Solar System. The combined volume of the four giant planets (Jupiter, Saturn, Uranus and Neptune) is just over a thousand

THE INTERIOR OF THE EARTH

You might think that all we can know of the Earth's interior must be speculative, because it has not been possible to drill down any significant distance. Yet scientists can probe through the Earth, using geophysical techniques, and evaluate the physical properties of the unseen materials in the Earth's interior.

These techniques have been almost exclusively seismic in character, using either natural earthquakes or man-made explosions. Simultaneous observations are then made of the arrival of the sound waves at different parts of the Earth's surface. On average, the Earth's outer continental crust is about twenty to forty miles thick, whereas the oceanic crust is only three to five miles thick. The upper and lower mantle are about 590 and 1,180 miles thick respectively, and the outer core is about 1,310 miles thick. The inner core has a radius of about 840 miles.

The **crust** is composed of many diverse elements, and there is much variation in the chemical composition of the rocks from region to region. The younger marginal platforms adjacent to a continent consist mainly of sedimentary rocks derived from continued erosion of the continental surface. The platforms often form beds up to several miles in thickness. The oldest known rocks of the PreCambrian Shields generally consist of igneous rocks such as granite and highly metamorphosed gneiss rocks.

The **mantle** forms the major part of the Earth (80% by volume). It is thought to be largely composed of mixtures of magnesium and iron silicates.

There have been many speculations about the nature of the materials which make up the **inner and outer core**

materials. Most of the studies have pointed towards a liquid/solid metal core made up from iron or iron nickel, but these can only be inspired scientific guesses.

Partly fused peridotite
To depth about 590 miles; peridotite with high-density minerals

Upper mantle
To depth about 125 miles; various types of peridotite (olivine and pyroxenes)

Phase of chemical change
From basalt rocks to peridotite rocks

Crust
20–40 miles thick at the continents, 3–5 miles thick under the oceans; made of basalt silicate rocks

Crust

Upper mantle

Lower mantle
To depth about 1,800 miles;
peridotite with high-density (mainly
silicate) minerals: periclase MgO,
stishovite SiO_2

Outer core
To depth about 3,200 miles; liquid,
metal-like substance (iron and nickel
with some less-dense material such
as silicone)

Inner core
To depth 3,960 miles; solid metallic
(could be iron and nickel but no
evidence)

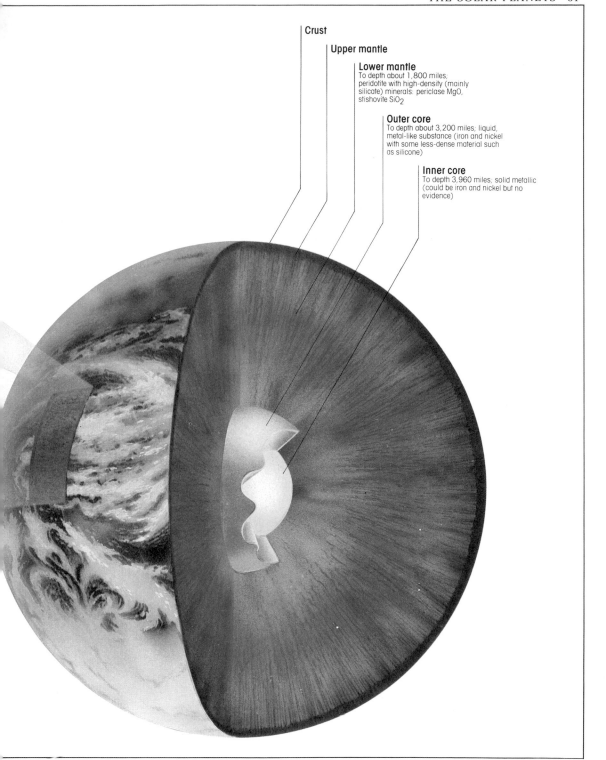

times that of the four inner planets, but their combined mass is only a little more than 200 times greater.

The outermost planet so far identified is **Pluto**. It has a very elliptical orbit and grazes the extra-planetary fringes of the Solar System.

Mankind has long been curious about the origin, history and nature of the Sun and its planets, and much scientific progress has been made towards answering some of the questions. But in recent years these inquiries have been helped by the contemporary technology of space travel. For the cosmologist and cosmochemist the recent space ventures have illuminated their study of the planets and their moons. But the world-wide popular interest in them has mainly been over the specific question 'Is there evidence for life on the planets or on the Moon?' This is the question we all want answered. Let us consider the planets one by one:

● **Mercury**, the Sun's closest neighbour, is the Solar System's smallest member. Looking at the planet through a telescope can be very frustrating: because its orbit lies inside that of the Earth, its back is turned when it is closest to us.

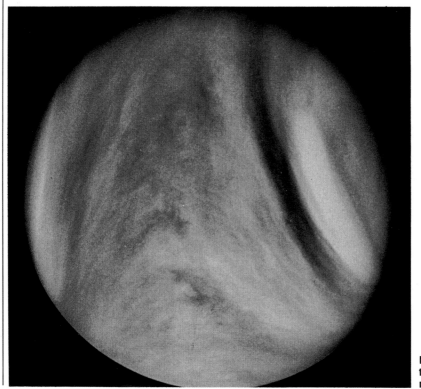

NASA's Pioneer 2 spacecraft took this photo of Venus, Earth's nearest planetary neighbour.

When the day-time side does face the Earth, the planet is behind the Sun. Our knowledge of Mercury increased many, many times through the success of the Mariner 10 mission in 1974.

At midnight on Mercury it is very cold: temperatures fall to around 170°C below zero. Because the planet has no atmosphere, heat accumulated by the rock surface during the day moves away into space soon after sunset. Early in the Mercurian morning, the temperature climbs back up over 430°C. With these extremes of temperature and with no atmosphere, it is very unlikely that life does or ever did exist on Mercury.

● **Venus** is the Earth's twin in terms of size, and it is also our nearest planetary neighbour. In December 1962, remote-sensing equipment aboard the US spacecraft Mariner 2 flew within 22,000 miles of the planet; it recorded a surface temperature of about 430°C — as high as Mercury's. In 1967 and 1969 the Soviet Venera spacecraft encountered correspondingly high temperatures in the 'greenhouse' atmosphere of carbon dioxide. In December 1970, Venera 7

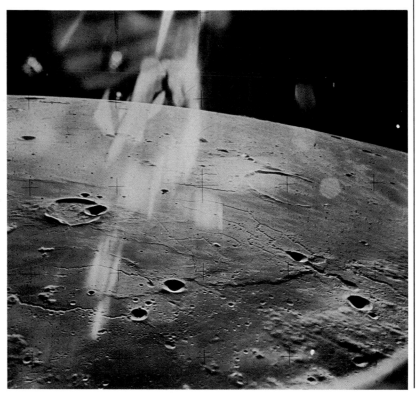

This photograph of the Moon's surface was taken from the Apollo 15 command module. The sunflares and reflections are caused by the glare of the Sun's light.

reported a direct surface temperature of 475°C! The Venus atmosphere is extremely dense and the surface pressure is about ninety times greater than it is on Earth. A few decades ago there were ideas that Venus had a jungle-like environment, or that it was an endless desert, or that it was completely submerged by oceans. These are only possible in the realms of science fiction.

● **The Earth**, which is the largest of the 'terrestrial' planets, cooled down between 4,500 and 4,000 million years ago and formed a crust. Today the crust is discontinuous, being made of a number of closely-fitting plates like a children's jigsaw. Each of these seven major plates has played a major role in the history of the Earth. (See *Plate Tectonics* and *Continental*

CONTINENTAL DRIFT

The jigsaw-puzzle fit of the Earth's continents, including Africa and South America, is impressive to the eye, but it is actually one of the lesser pieces of evidence to support continental drift.

Today the great majority of earth scientists accept as a fact that the present distribution of the continents results from the break-up and joining together of previous distributions of the continents. This process, known as 'Continental Drift', has probably been at work for a very long time, perhaps ever since the Earth first cooled down and formed its crust.

When reconstructing the picture of how the present continents were once joined together in one super-continent, scientists now use the detailed shapes of the outer continental shelves, which mark the boundary between the thin crust of the sea-floor and the thick crust of the continents. When this continental-margin jigsaw is pieced together, it gives a good picture of the ancient super-continent —

much closer than that given by the outlines of continents at sea-level.

Studies of continental geology suggest that the drifting fragments which we call continents were derived from the break-up of an ancient super-continent, *Pangaea*. Evidence for this theory has come mainly from fitting together continental margins and seeing how geological structures and formations can be fitted together and joined up across different continents. The distribution of certain plants and animals in the past, as well as ancient climatic zones, could not be satisfactorily explained without the idea of continental drift.

These studies reveal a very unfamiliar map of the past. Some continents now separated by wide oceans (Africa and South America, Africa and Australia) were once connected. And other land masses now close or connected (North Africa and Southern Europe, India and Asia) were far apart.

Using correctly-defined continental boundaries and applying modern computer modelling, it is now possible to get an almost perfect fit between Africa and South America, with the other continents fitting closely together into two super-continents: *Laurasia* in the Northern Hemisphere and *Gondwanaland* in the Southern Hemisphere.

The pattern of continental drift is known in detail for the last 200 million years, the period during which *Pangaea* broke up. The movements are also known in outline of continents which converged to form *Pangaea* 350 million years ago. PreCambrian continental drift has not yet been explained.

As the first super-continent, *Pangaea I*, broke up about 600 million years ago, the diversity of animals increased dramatically, until four new continents were formed around 450 million years ago. This faunal diversity continued until about 200 million years ago,

Drift.) The Earth's crust is seven-eighths covered by water which, with the atmosphere of oxygen and nitrogen, supports life. Surface temperatures range from 60°C above to 90°C below zero. The Earth has a single Moon. The Earth and the Moon are companions in Space. Although only the sixth largest in the Solar System, the Earth's Moon is by far the largest satellite in proportion to the dimensions of its mother planet (about a quarter of Earth's diameter and about one-eightieth of its mass)

Two main theories were proposed for the Moon's origin. The 'daughter' hypothesis suggests that the Moon spun off from the Earth, like a lump of clay flying off from a potter's wheel. The alternative theory is that the Moon formed separately at the same time and place as the Earth and is a

when these four continents re-combined to form *Pangaea II*. This second super-continent was relatively short-lived and appears to have broken up to form *Laurasia* and *Gondwanaland* separated by an ocean. These land masses gradually broke up and began to give the continents as we know them today, although in different geographical positions. Further break-up, some 135–65 million years ago (at the end of the Cretaceous Period) began to give the continents an appearance rather like that well-known picture of our modern globe.

These projections show the map of the world as it would have appeared at different stages of Continental Drift. They represent the Earth successively 200 million, 180 million, 135 million and 65 million yeas ago. You can see the stages by which our present distribution of land masses was reached.

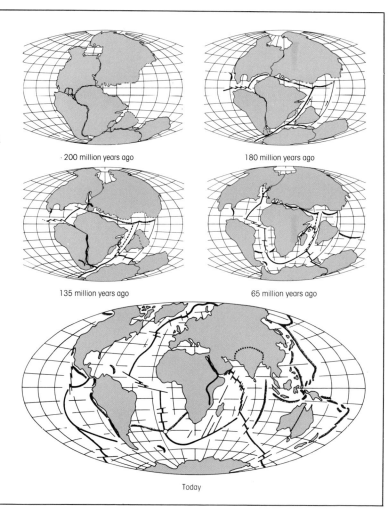

200 million years ago

180 million years ago

135 million years ago

65 million years ago

Today

PLATE TECTONICS

The plates are continuously in motion with respect both to each other and to the Earth's axis of rotation. Virtually all major geological activity — volcanoes, earthquakes, mountain building and other tectonic activity — is localized around **plate margins**. It is associated with differential motion between adjacent plates.

There are no gaps between the plates, nor is there significant overlapping. The plates move principally by easing themselves against each other, or if they move towards each other, then one plate dips and its material passes under the edge of the other to re-enter the Earth's interior in what is called a **trench**.

An individual plate is by no means a permanent feature of the Earth's crust. A plate without continental crust can entirely disappear down a trench. Changes and major activity are especially likely whenever three plates meet, as each of them is moving in a different way (a triple junction) causing them to shift their positions.

A large number of countries and cities are situated on **active faults** caused by plate edges. Perhaps the best known is the San Andreas fault which runs through California. This major fault, together with an adjoining network of lesser faults occurring on a fault-line near San Francisco, is the source of the ground movements that cause California's earthquakes. The San Andreas is a **transform fault** at the boundary where

two huge plates (the American and Pacific plates) slide past each other and occasionally pull apart or press together.

A most obvious example of plate tectonics is mountain building. The theory suggests that every great mountain chain in Planet Earth has been produced by plate movements and usually by the destruction of an ocean. There is clear evidence of recent plate movement that tells us that sub-continent India has collided with Asia during the past 50 million years and made the Himalaya Mountain Range. The ocean that lay between India and Asia has vanished, leaving a chain of mountains lying deep inside an enlarged continent.

The Earth that life developed on, and the continents that first contained life, would have looked very different from the modern Earth when viewed from space. Planet Earth has not always had the same arrangement of continents and oceans.

Plate tectonic movements, continental drift, sea-floor spreading, mountain building, volcanoes and earthquakes have changed the face of the Earth repeatedly, and often with important consequences to the origin, development and extinction of life on Earth.

Professor Alfred Wegener, a German astronomer-turned-meteorologist, is widely regarded as the father of the concept of Plate Tectonics; he published his ideas as long ago as 1915. But the main research on 'Plate Theory' began in the

early 1960s when the late Harry Hess of Princeton University published his circumstantial evidence and brilliant geological insight into sea-floor spreading. Later work has confirmed these ideas and work on ocean basins has shown that plates, movement, ocean ridges and sea-floor spreading were not just speculation, but could be predicted. This science of 'New Global Tectonics', now usually called 'Plate Tectonics', affects virtually all disciplines of Earth science.

Tectonics means 'building' or 'construction'. Plate tectonics is the construction of the Earth's geological features by the interaction of the seven major plates of the Earth's crust. The rocky surface and the outermost thirty to forty miles of the modern Earth's surface consist almost entirely of cool rock that has solidified and is considered to be rigid. This outer shell of the Earth is 'cracked' into a number of plates, which are able to move about depending on the geological and geophysical conditions of the Earth.

Mountains are built as tectonic plates collide. The Himalayas, topped by the majestic Everest, are 'fold mountains', formed as the once-separate India collided and joined up with Asia.

kind of 'sister' planet. A less-accepted alternative theory is that the Moon formed somewhere else and was captured by gravity into the Earth's orbit; this is called the 'wife' hypothesis.

The subsequent history of the Moon has been fitted together from evidence brought back to Earth from six successful Apollo missions between July 1969 and December 1972. The missions brought back lunar rocks for analysis and study, and also left behind instruments on the lunar surface. From photographs taken by the Apollo missions, and information obtained from unmanned American Lunar Orbitor and Surveyor and Soviet Luna spacecraft, we have received a wonderful pictorial lunar atlas. But even more importantly they have given us much important scientific and technological information.

This photograph, taken by Voyager 1 on 1 February 1979, shows the planet Jupiter with three of its moons — Io, Europa and Callistro. Jupiter is the largest and nearest to the Sun of the great 'outer planets', which are made of gas.

This astronaut from the Apollo 12 moon mission is carrying out scientific experiments. He is seen against a panoramic view of the Mare Cognitum area of the Moon's surface.

● **Mars**, the 'red planet', has always attracted a lot of attention compared with the other planets, because it was considered a possible habitat for life. Mars is on average one-and-a-half times farther from the Sun than is the Earth.

THE MOON

Telescopes, lunar probes and manned landings have told scientists much about the Moon. We now know its gravity is low, it lacks water and air, its days are very hot (130°C) and its nights intensely cold (minus 200°C). No animals or plants could naturally survive on the Moon.

The Moon originated about 4,600 million years ago along with Planet Earth and the rest of the planets and their moons. The smallness of its mass and the relatively low value of its surface escape-velocity have several important consequences. The most important is the complete absence of any atmosphere or protective gaseous mantle above the Moon's surface. Because it possesses no detectable atmosphere by virtue of its low gravitational field, and because of its relatively high daytime temperature, the Moon cannot maintain liquid on its surface. Hence no water or ice could be present almost anywhere on the Moon's exposed surface except possibly for very short intervals of time.

The surface of the Moon is thus bone-dry, and probably has been since its origin. The lunar landscape has turned out to be what was always predicted — arid, stark and lifeless.

Facing page
Astronaut Buzz Aldrin at Tranquillity Base on the Moon, in July 1969. Neil Armstrong is reflected in Aldrin's visor.

The Moon, Earth's nearest planetary neighbour and the only extra-terrestrial body visited by a manned spacecraft.

It has a radius about half that of our planet and a mass of approximately one-ninth. The Martian year is nearly twice as long as the terrestrial year.

The present atmosphere of Mars is more than a hundred times thinner than that of the Earth. And the surface temperature, apart from a few daylight hours at equatorial latitudes, is well below zero, so that any water would be frozen. Given these facts, Martian conditions are currently considered hostile to life.

● With the exception of our most distant solar neighbour Pluto, the **asteroids** will be the only bodies in our Solar System left unexplored by spacecraft at the end of the 1980s. Although their total mass is small (about one per cent of the Earth's mass), the asteroids are of great importance to our understanding of the Solar System. Asteroids mark the transition from rocky, 'Earth-like' planets to the huge gas giants of the outer regions, and if current theories of the Solar System's origin are correct, they may offer a unique window on the system's early history.

The asteroids are small, rocky bodies, most of which orbit between Mars and Jupiter. The radii of the asteroid orbits are about two to four times as big as the radius of the Earth's orbit. The larger asteroids may have metal-rich cores. Others may be rocky or metallic fragments, the probable products of collisions between the members of an earlier generation of perhaps thirty asteroids. Many asteroids are very dark

The planet Saturn, the ringed planet, with some of its ten moons. This true-colour view was taken by Voyager 2 in 1981, from a distance of 8.1 million miles.

VIKING EXPLORATION ON MARS

Mariner spacecraft surveyed the whole of the Martian surface photographically on separate visits to the planet during the 1960s and early 1970s, providing high-resolution pictures. A major spacecraft achievement was the Viking 1 and Viking 2 probes. Launched from Earth orbit in August 1975, Viking 1 travelled a 4,375 million-mile interplanetary spiral to reach Martian orbit the following June. On 20 July, a car-sized 'lander' dropped from the mother spacecraft. It landed on the rock-strewn surface at *Chryse Planitia*, about 1,250 miles north of the eastern end of the Mariner canyon system. Viking 2 landed on 3 September 1976. Both spacecraft were to search for evidence of life.

Both spacecraft were equipped to carry out biological experiments in situ on samples of Martian soil. The presumption was that any micro-organism which might be present in the soil would have metabolic processes similar to those of terrigenous micro-organisms. A series of experiments was carried out by the lander spaceship: the soil was treated with various nutrients and any expelled gases were analyzed.

The results of these experiments turned out to be rather confusing and taken together the results were, in the words of biology team-leader Harold Klein, 'somewhat controversial'. In some experiments something in the Martian soil had consumed carbon dioxide, 'eaten' nutrient and produced oxygen on contact with water vapour, all of which could be regarded as indicating some type of biological process. But other experiments showed there to be no micro-organisms or even any of their remains or detectable amounts of organic compounds. Although some of these results were marginally consistent with the existence of Martian micro-organisms, scientists interpret these findings as very far from proven.

A chemical explanation has been suggested for the exotic behaviour of the Martian soil samples. The super-oxidant property of the soil is explained by the dissociation of water into free oxygen and hydrogen by solar ultra-violet light, with light hydrogen escaping into space, so leaving an excess of oxygen on Mars. This would mean that the reactions observed are inorganic phenomena, not biochemical ones.

The lander module from a Viking space probe was equipped to take samples of soil from the surface of Mars and carry out biological tests on it.

bodies. These, it has been suggested, may be coated with a layer of the oldest material in the Solar System, originating when the gas-and-dust cloud surrounding the young Sun started to cool.

Beyond the Earth, Mars and the asteroids are the five planets farthest from the Sun. The first four giants are far larger than the Earth; compared with them Pluto is a dwarf.

● **Jupiter** is twice as massive as the other planets put together. It spins faster and has more moons than any of the rest. Beneath its cloudy surface this giant may be mainly made of liquid hydrogen.

● **Saturn** is the second largest planet and is mainly made of three gases: hydrogen, helium and methane. Flat rings encircle its equator and make the planet the most impressive and beautiful to look at in the Solar System.

● **Uranus** is about four times as far across as the Earth. Like Jupiter and Saturn, it is mainly made of gases. It is intensely cold. As recently as 1977, Uranus was discovered to have a ring system like Saturn's, only very much smaller and fainter.

● **Neptune** is very like Uranus in its size, make-up and intensely low temperature. But it has only two known moons, compared with the five on Uranus.

● **Pluto** is the remotest planet. It is nearly forty times farther from the Sun than the Earth is. Pluto may be even smaller than Mercury.

As we move farther out from the Sun into the Solar System, the prospects for life on the planets rapidly recede. The possibility of life ever having existed on Jupiter and Saturn is almost zero. The chemical and physical conditions on the outermost planets, Uranus and Neptune, make the presence of life as we know it impossible. The most distant solar planet, Pluto, remains relatively unknown, but life is almost certainly not present on this planet.

So we have drawn a blank. Life may possibly exist in the planetary systems of other stars, but in our Solar System it seems Earth is the only planet supporting life.

The Early Earth and the Origins of Life

O f all the mysteries of biology, unquestionably the most baffling is how life originated on Earth. Molecular biologists, palaeobotanists, geologists and traditional biologists have gradually moved backwards in time, studying and testing their evolutionary ladder. At last they came to the period in the Earth's history where they had to stop and ask the question, 'How and when did life arise on Planet Earth?'

So far as we can tell from the record of terrestrial rocks and from measurements of their radioactivity — together with comparable radiodating of meteorites — the Earth began its independent existence within our planetary system about 4,500 million years ago. The earliest evidence for life so far identified is found in sedimentary rocks formed more than 3,500 million (possibly up to 3,800 million) years ago. However, as H. G. Wells pointed out, the record in the rocks is no more a complete record of life in the past than the books of a bank are a record of everybody in the neighbourhood. A particular morphological living system detected in ancient sediments may not necessarily bear much relationship to the original organism from which it was derived. Many, many millions of early life-forms and their more complex but soft-bodied descendants may have lived and flourished and passed away without leaving any remnant to be preserved in the fossil record.

If some extra-terrestrial space expedition to Planet Earth

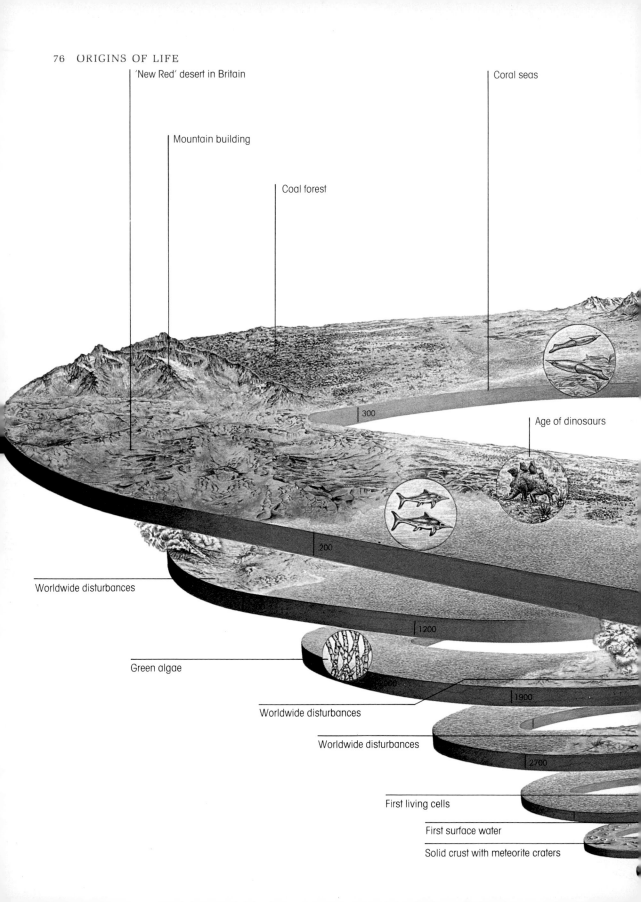

'New Red' desert in Britain

Coral seas

Mountain building

Coal forest

Age of dinosaurs

300

200

1200

Worldwide disturbances

Green algae

1900

Worldwide disturbances

Worldwide disturbances

2700

First living cells

First surface water

Solid crust with meteorite craters

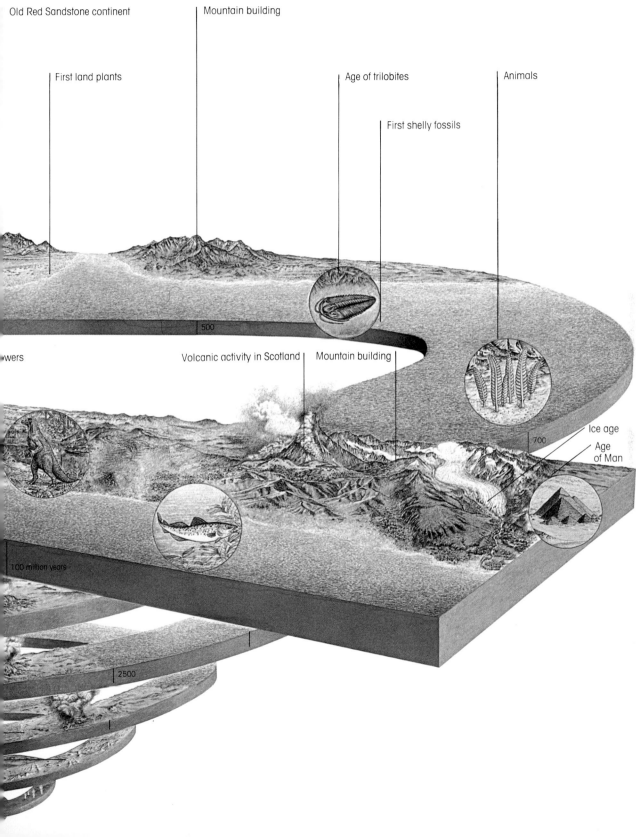

Old Red Sandstone continent

Mountain building

First land plants

Age of trilobites

Animals

First shelly fossils

500

wers

Volcanic activity in Scotland

Mountain building

700

Ice age

Age of Man

100 million years

2500

were to take place, and if their 'government' were to restrict their budget so that only two samples could be collected to determine the history of life on Earth, their geological advisers would probably suggest collecting samples of stromatolites or black cherts to represent the first few thousand million years of biological history. Imagine, then, that the visiting spacecraft first sample rocks in Southern Africa and collect stromatolites and black cherts from ancient rocks, both more than 2,700 million years old. Next they travel across the Indian Ocean to Shark Bay in Australia (or some similar arid shore), where they collect a sample of extant modern life from the intertidal zone. The spacecraft return home and the samples are analysed in their laboratories. They will find that the cyanobacteria organisms are the same in the ancient and modern rocks. This will lead them to conclude that there has been no evolutionary progress for at least 2,700 million years and that Planet Earth is still in the age of primitive bacteria.

We must be aware of the biases that can result from selective preservation. Not surprisingly, therefore, the jigsaw of the origin of life on Planet Earth and the course of early evolution still has a number of scientific pieces to be added!

There have been many suggestions, theories and suppositions about how life began on Earth. But four main theories have developed through history and become established in our thoughts and discussions:

● **Supernatural creation** is the belief that God the Creator uniquely stepped into his creation of the Universe, and created life on the Earth by a sovereign act.

● **Spontaneous generation** was a very popular idea, even as late as the nineteenth century. It was based on the (wrong) observation that all kinds of maggots and insects appear from the damp earth, rotting meat and so on. This led people to suppose that dead matter can somehow produce life. In the late nineteenth century, however, the work of Louis Pasteur finally put this idea to rest, when he showed that properly sterile matter never produces life. In his Presidential Address to the British Association in 1871, Lord Kelvin said, 'Dead matter cannot become living without coming under the influence of matter previously alive. This seems to be as sure a teaching of science as the law of gravitation.' Attitudes at that time took the line that, since spontaneous generation was impossible, therefore life must be eternal.

● **Universal life** is a theory based on the assumption that life, like matter, has always existed somewhere in the Universe. This idea goes back to the Greek Anaxagora in the fifth-century BC. It was re-introduced during the early 1900s by the Swedish chemist Svante Arrhenius, who introduced the revolutionary idea known as the 'Panspermia Hypothesis', suggesting that life was seeded on Earth by primitive organisms carried by meteorites and/or comets from outside the Solar System. (See *Life from Space*.) This discussion was revived in the 1970s by two British professors, Fred Hoyle and Chandra Wickramasinghe, who proposed that life arrives in the form of bacteria embedded in cometary material. They think it is still arriving. Hoyle's claim that the interstellar clouds contain bacteria is not accepted by scientists in general and astronomers in particular.

● **Abiogenic synthesis** ('abiogenic' means 'non-biological') is a modern scientific concept of how life began on Earth. It follows from proposals by the Russian Academician Aleksandr Oparin in 1924 and Englishman J. B. S. Haldane in 1928. They put forward the novel idea that the Earth's primitive atmosphere was reducing (that is, it did not contain free oxygen). They suggested that it contained ammonia, water, carbon monoxide and so on. Their hypotheses also included the addition of energy sources (such as lightning and ultra-violet light) to the primitive atmosphere. These might have caused the synthesis of biologically-interesting molecules in a primitive ocean.

Professor J. B. S. Haldane, the British scientist who first proposed the Chemical Evolution Theory of life's origins.

These theories are not necessarily mutually exclusive. It is possible that two or more may have acted together in the processes leading to the origin of life on Earth. God may have created life using abiogenic synthesis, for example.

Discoveries during the last decades in geochemistry and palaeochemistry provide the means to examine these concepts scientifically. Scientists are now forming views on the nature and significance of organic matter in the Earth's ancient rocks and in extra-terrestrial objects, and the role of such matter in the origin and development of living systems.

It seems reasonably clear that the surface of the Earth some 4,000 million years ago was so hot that no life could exist, nor for that matter could one expect even complex macromolecules (such as proteins and nucleic acids) to exist in any form remotely resembling that required for life. So at

DEFINING LIFE

In all discussions on how life began, a major problem is how to define 'life'. Life is an elusive thing to define fully and completely; each time a definition is approached it slips away, often because there always seems to be an exception to any particular rule. Simple lists — a 'living' object moves, grows, eats, excretes, reproduces and so on — do not suffice, because there are always significant exceptions to the list: organisms that do not apply the basic definition.

Aristotle believed that nowhere on a line drawn from the smallest atom (hydrogen) to the most complex living thing (man) was it possible to say where non-life ended and life began. Many scientists of today would agree with him, suggesting that matter simply differs according to its degree of organization.

Recent advances in chemistry and biochemistry have led to

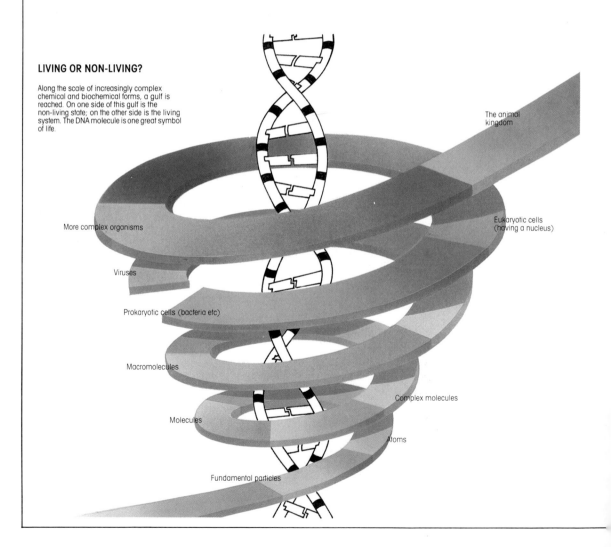

LIVING OR NON-LIVING?

Along the scale of increasingly complex chemical and biochemical forms, a gulf is reached. On one side of this gulf is the non-living state; on the other side is the living system. The DNA molecule is one great symbol of life.

The animal kingdom

More complex organisms

Eukaryotic cells (having a nucleus)

Viruses

Prokaryotic cells (bacteria etc)

Macromolecules

Complex molecules

Molecules

Atoms

Fundamental particles

definitions of the living system, but these quite rigid definitions of life are couched in purely physical and chemical terms. Some basic characteristics have been combined to give useful definitions:

'A system that everyone would presumably accept as living can be defined as a metabolizing system that reproduces itself, mutates, and reproduces mutations.'

'Wherever living things exist, they are likely to be carbon-based, dependent on a supply of liquid water and contingent on either the absence of more than minute quantities of free oxygen or the presence of suitable oxygen-mediating enzyme systems.'

'Life is the structure-replication of enzymes ensured by exactly-reproduced nucleic acid molecules.'

'Life has the capacity for self-repair.'

But the rigidity of some of these detailed molecular-biological definitions is their weakness, since they fail to describe the system and only describe one or more of its functions.

In a paper I wrote with G. Shaw in 1978, we suggested that Aristotle's continuum from the hydrogen atom to the human living system is wrong. There are obvious chemical and physical relationships existing between the various elementary particles that make up the atom, and from the hydrogen atom to the nucleoproteins in the spectra of materials, including the viruses. Equally, there appears to be an established evolutionary sequence from prokaryotic cells (without a clearly-defined nucleus), through various types of eukaryotic cells (which have one), to the highly-differentiated structures represented by the human living system. However, between the nucleoprotein structures and the simplest of prokaryotic and eukaryotic cells there is an enormous gulf. This gulf represents a basis for a definition of life: on one side of the gulf there is the non-living state, and on the other side the living system.

This scientific definition of life may have advantages over many other narrower definitions. The most useful description of the living system is that it is like a biological machine. Everything known about the living system, taking into account recent advances in chemistry and biochemistry, leads one to the conclusion that the living system is very closely analogous to the machine. No originality is claimed for this concept, which is frequently expressed and almost regarded as axiomatic in modern molecular biology.

However, there is a difference between regarding the living system as a machine with qualifications, as is normally implied in molecular-biological definitions, and regarding it as a machine pure and simple. Unlike a motor car, the living system can reproduce itself. If we could look at a car factory in which all the operatives were invisible (including the people who had programmed the computers), we would see what appeared to be a living system creating and re-creating itself over and over again. Then suddenly this machine mutates (a new model) to a form which perhaps is more powerful or safer or possessing other useful features which apparently seem to arise by some process of natural selection. Certainly, in one way, a driven motor car is every bit as alive as the person driving, since the living system can clearly extend from the driver to the driven in a quite easily-understood manner. Only when contact with the living system is removed do we see the motor car as a dead, inanimate thing: a sort of mechanical object that might on some future occasion spring to life.

The analogy between the living cell and the motor car can usefully illustrate the living system. The living cell, in spite of its intricacies, can be seen as a collection of physical and chemical nuts and bolts, which could well have been created by a supernatural creator and controlled and driven by him ever since.

some time there must have been a period when life arose on a sterile planet. There are several ways in which this might have occurred.

Our present knowledge is derived from five main study areas:

- **Studies of the formation of the Earth's biosphere, atmosphere and hydrosphere;**
- **Laboratory experiments designed to test abiogenic-synthesis theories;**
- **PreCambrian rocks and early life;**
- **Attempts to explain the presence of organic compounds in carbonaceous chondrites (a type of meteorite), and on the other solar planets and the Moon;**
- **The study of organic compounds in interstellar space.**

Chance or Purpose?

I magine a planet made of nothing but parts of television sets. It has been proposed that in the fullness of time (1,000 million years), physical forces and the restless motion of the Earth would randomly assemble at least one complete working television. The reasoning behind the abiogenic origin of life on Earth is similar to this idea of the 'chance manufacture' of our television.

It is suggested that the countless number and variety of random encounters between individual molecular components and precursors of life may eventually have resulted in a chance association of parts which together could perform a life-like task. They may, for example, have carried out simple photosynthesis (collecting sunlight), and used this energy to execute simple 'biochemical' reactions that would otherwise be impossible or forbidden by the law of biophysics.

Proteins and/or nucleic acids are fundamental macromolecules for basic biochemical reactions. These molecules must have been the early precursors of the living system.

Proteins are long molecules with large molecular weights, composed of amino-acids linked together in very specific order and structure. Some proteins have as few as fifty or so amino-acids, others have tens of thousands. But each is only that protein rather than another because of the exact order of amino-acids in its molecular chain.

Getting the order of the amino-acids correct in the protein molecule is the job of the gene for that protein. The gene, a length of DNA (deoxyribosenucleic acid) which carries the genetic code for that protein, is coded in such a way that three DNA 'letters' are needed for each amino-acid. In other words, it takes a length of DNA 300 letters long to carry the genetic instructions for a protein of a hundred amino-acids.

To appreciate the complexity of this fundamental but essential biochemical process, and the possibility of it occurring by chance to give the first primitive protein, we can make use of a well-known illustration. Imagine a tireless monkey trying to type out the 300-letter word of his single 300-letter gene, remembering that there are four different letters. The odds against getting the first letter correct are 4–1. But the odds against getting the first *two* letters correct are 16–1 (4×4). For the first *three* letters, the odds are multiplied by four again to give 64–1. And so on for the remaining 277 letters of the code for the gene. Since some DNA letters can be replaced by alternatives and still give the correct amino-acid, in the end it turns out that for every attempt to spell out the complete gene-word code, the odds against getting it correct are approximately 1 in 10^{130} (10 followed by 130 zeros).

The monkey could of course type out the correct word-code at the first attempt or the ten thousandth or any time after that, if he goes on for long enough. But what these odds are really illustrating is not to expect it, even if you could have a thousand million monkeys on each square inch of the Earth's surface typing at the rate of a thousand million words per second for a thousand million years. Even with this high-powered monkey set-up, the odds against the correct gene-sequence turning up are still 1 in 10^{80}. This illustration considers just one gene with a mere 300-letter code. But if we try to extend our reasoning to DNA present in a *simple* virus, which could have a 20,000-letter code, the average simple bacterium with about 4.5 million-letter code and the genetic information of a human being with roughly 5,000 million, the complexity and statistical chances of forming such genes approaches infinity; even mathematicians must consider it unlikely.

This generalized proposition — that processes of chance and natural law led to living organisms emerging on Earth from the relatively simple organic molecules in 'primordial soups' — is valid only if there is a finite probability of the correct assembly of molecules occurring within the time-scale envisaged. Here there is another great problem. In the above example for a relatively small protein of 100 amino-

The coloured spheres on this model of the complex DNA molecule represent atoms of carbon, hydrogen, oxygen, sulphur and phosphorus. DNA carries hereditary information in cells, so making it possible for cells to replicate themselves.

acids, selection of this correct sequence had to be made by chance from 10^{130} alternative choices. The operation of pure chance would mean that within a maximum of about 500 million years (or somewhat less), the organic molecules in the 'primordial soup' might have to undergo 10^{130} trial assemblies to hit on the correct sequence. The probability of such a chance occurrence leading to the formation of one of the smallest protein molecules is unimaginably small. Within the boundary conditions of time and space which we are considering, it is effectively zero.

HAEMOGLOBIN AND CHLOROPHYLL

An essential component of the human and other vertebrates is the protein haemoglobin. Contained in red blood cells, haemoglobin serves as the oxygen-carrier in blood. It is made up of four polypeptide chains which are held together by co-valent interactions. Haemoglobin is only able to bind and carry oxygen because of the presence of a non-polypeptide unit, namely a *heme* group. (The *heme* gives haemoglobin its distinctive red colour.) Haemoglobin is made up of amino-acids joined together in twisted chains to make its bio-chemically-unique structure. Each amino-acid is uniquely linked together in a specific sequence of 574 components forming four principal chains which are interwoven around each other to form the complex 3-D haemoglobin molecule.

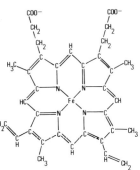

The chemical structure of the heme molecule, an important compound in haemoglobin, which is the red component in blood. Heme consists of an organic part and an iron atom.

A healthy human has about 60 million million million (6×10^{21}) identical haemoglobin molecules of which 400 million million (4×10^{14}) are destroyed every second and replaced by identical molecules synthesized by the body. The human cells do not synthesize such compounds by random chemical processes, but by specific enzyme-controlled biochemical reactions.

It seems to me that even if there were primeval soups of chemicals on the primitive Earth (or in the oceans), no random/chance abiogenic process operating at that time could possibly have given molecules as complex as haemo-globin, nor any other complex protein molecule such as chlorophyll — let alone a living cell.

The fundamental driving force in the biological cycle on Planet Earth is that all life runs on light — sunlight received from our Sun. The mechanism by which plants absorb and use this sunlight-energy is photosynthesis, and this process depends on a substance known as chlorophyll. (The chemical structures of haemoglobin and chlorophyll are similar.) The

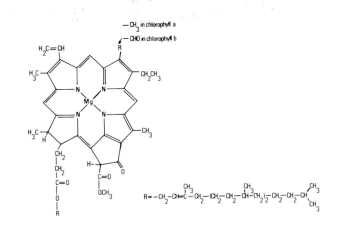

The chemical structure of the chlorophyll molecule. Chlorophyll is a main chemical in green plants. It has an important role in the complex biological process of photosynthesis.

chemistry of photosynthesis is very complex and involves a large number of interlocked, interwoven biochemical reactions. Until the highly-specialized chemical structures of chlorophyll had developed and the complex process of photosynthesis had begun to operate, it was not possible for plants to evolve.

The emergence of photosynthesis in plants had a tremendous impact, not just on plant evolution and development, but also on the whole history of Planet Earth. It is considered that photosynthesis changed the Earth's atmosphere from one based, probably, on nitrogen to our current atmosphere based on oxygen. When this fundamental change to an oxygen atmosphere took place is a matter of much discussion. But until that change happened, the environment to support life as we know it today could not come into existence. It is still to be demonstrated how these essential molecules, such as haemoglobin, chlorophyll and other proteins and nucleic acids were formed. But even if we were to allow a primeval soup to have existed for the full history of the Earth (4,000–4,500 million years), complex proteins and nucleic-acid molecules could never have been produced by random, chance interactions.

However, here are you and I on Earth today. And the evidence of the fossil record shows that some sequence of events of almost zero probability did take place over 3,500 million years ago. Before the event, the chances that it would occur were exceedingly small. What is more, from our understanding of the possible processes leading to the origin of life and the critical part played by living organisms in the development processes, the transition from non-living to living matter probably occurred only once and could have occurred only once.

The origin of life was an almost utterly improbable event with almost impossible odds against a chance happening. But life did originate. So was it by chance? Or was it by design and control?

Biosphere, Atmosphere and Hydrosphere

T he Earth, like the rest of the Solar System, came into existence 4,500–4,700 million years ago. It is considered to have had no significant atmosphere then, but an environment rich in hydrogen may have been present. The Earth's outer crust, the initial atmosphere and hydrosphere (the oceans) were probably formed by the same separating and degassing processes of the mantle about 4,000 million years ago.

The initial 'primordial' hydrogen environment quickly dispersed by diffusion, and then a secondary atmosphere was introduced by volcanic activity. From occasional outbursts of solid and fluid matter and the release of gases and vapours, an atmosphere formed that had reducing properties — contained no free oxygen. Water vapour (steam) is the most abundant constituent (97%) of the volcanic gases and it is estimated that the quantity of water and gases released over the last 4,000 million years more than accounts for the volume of the oceans and for the nitrogen and other constituents in the present environment.

How did oxygen enter the atmosphere? Oxygen originated and increased in the atmosphere by dissociation of water under the action of solar energy. Several mechanisms for the dissociation of water have been proposed and the three most important are:

● **Photosynthesis**, a process by which chlorophyll-bearing plants synthesize food in the form of carbohydrate,

Volcanoes are among both the most disturbing and the most beautiful of natural events. Their geological after-effects are also interesting, as seen in the Cretor Moon Natural Monument in Idaho, United States (below). The theory of chemical evolution suggests that the energy and the superhot environment of volcanoes could have produced the atmosphere and conditions for the synthesis of molecules needed for life.

from carbon dioxide and water in the presence of light. A by-product of this process is oxygen.

● **Photolysis**, sometimes called 'photochemical decomposition', is a process in which compounds are broken down by ultra-violet light or very strong visible light of short wavelength. Water can be broken down into hydrogen and oxygen by photolysis.

● **Oxidation** of elements of variable valence.

The major process of dissociation is photosynthesis. Like any set of chemical reactions, photosynthesis does not produce a net change in oxidation. Oxygen is produced, except in bacterial photosynthesis, together with a quantity of reduced carbon. Almost all the oxygen is eventually used to oxidize this reduced carbon, and the only net gain in oxygen levels equals the amount of reduced carbon-material removed from the biogeochemical cycle and buried as sedimentary organic matter before it can be oxidized. Thus an important mechanism for fixing organic carbon in the Earth's crust and releasing free oxygen into the atmosphere was the accumulation of sedimentary organic matter that escaped oxidation by being deposited. These processes have been active since the first presence of photosynthetic organisms in the early PreCambrian period.

Geochemical studies and models have been made that show that sedimentary organic matter, close to eighty per cent (by weight) of the current terrestrial volume, already existed over 3,400 million years ago. This large terrestrial reservoir of PreCambrian sedimentary organic matter suggests that oxygen, derived from photosynthetic plant processes, was present in the very ancient rocks older than 3,400 million years.

The oxygen now in the atmosphere, together with oxygen combined with a wide range of other elements in the Earth's crust, is probably mainly if not wholly biological in origin. As the content of free oxygen in the hydrosphere and atmosphere increased, living systems evolved suitable oxygen-mediating enzymes within their cells and developed

The ancient equivalents of these modern blue-green algae are thought to have been some of the earliest life-forms, flourishing in the Early PreCambrian Age over 3,500 million years ago.

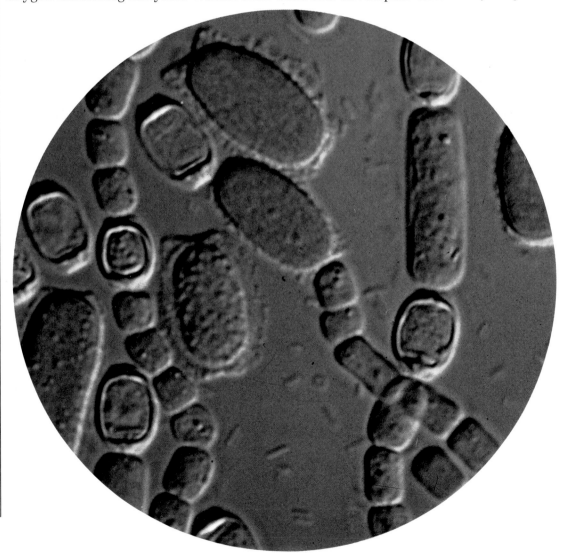

into more elaborate organisms with more sophisticated metabolic processes. The result of these and other processes is an intimate evolutionary interaction between the biosphere (living environment), atmosphere, hydrosphere and lithosphere (solid earth). The hydrosphere, which was probably formed soon after 4,000 million years ago, was the main regulator of the cycle of organic matter in the biosphere and sediments.

The timing of the build-up of oxygen in the atmosphere and hydrosphere is critical for the emergence of life, since many weathering and diagenetic processes involve oxygen-reducing reactions. The early presence of living systems on the Earth, which has now been established, implies that geological processes of sedimentation, weathering and diagenesis may have been influenced by organic activity from the earliest times.

The origin, composition and evolution of early Pre-Cambrian atmospheres have often been discussed. Conclusions are normally based on the following indirect evidence:

- **The primordial or primitive atmosphere was formed around the hydrogen-dominated solar nebula;**
- **The secondary atmosphere was produced from gases released from volcanoes;**
- **Abiotic synthesis of simple and increasingly complex organic chemicals requires a reducing or at least neutral atmosphere.**

Chemical evolution theory and the origin of life on Earth is based on the abiotic synthesis of organic molecules. It requires that the early Earth's atmosphere was reducing and contained free hydrogen. So, because abiotic synthesis of organic compounds theoretically requires an atmosphere without oxygen, the idea that a reducing atmosphere existed throughout the early PreCambrian has been built up. Not until recently was it questioned or tested by geochemical examination.

THE CHEMICALS OF LIFE

Proteins, nucleic acids and polysaccharides are the central, universal units of all living systems. They contain in their large molecular structures the whole complex variety of functions needed for cells to build and operate. What are they exactly?

We cannot go into much detail, but it will help here to describe these vitally important compounds, together with their respective building-blocks — **amino-acids** for proteins, **nucleotides** for nucleic acids, and **monosaccharides** for polysaccharides. There is also a large and somewhat ill-defined group of compounds called **lipids**.

Simple cells contain many types of chemicals, both organic (containing carbon) and inorganic. The least complex are simple substances like ammonia and common salt. But it is the compounds of larger molecular weight that really concern us.

AMINO-ACIDS AND PROTEINS
Proteins are naturally occurring 'polymers' (associations of units) of alpha-amino-acids.

Amino-acids consist of:
● an amino group;
● a carboxyl group;
● a hydrogen atom;
● a variable side-chain **group** bonded to a carbon atom, which is called the alpha-carbon.

Proteins play important roles in virtually all biological processes, but the most vital of all belongs to those proteins known as enzymes.

Enzymes are specific protein macromolecules which are the catalysts for nearly all the biochemical reactions which take place in cells. They usually enhance chemical reaction rates in living systems by at least a millionfold. More than a thousand enzymes have been named, and many more have been isolated. All known enzymes are proteins and they play the unique role of determining what pattern is taken by each biochemical reaction in a living system.

Proteins have a number of other important biological functions. For example:
● **Many small molecules and ions are transported by specific proteins;**
● **Proteins are the major component of muscle;**

● **The high tensile strength of hair, silk and bone is due to a protein — collagen;**
● **Antibodies are highly specific proteins which give immune protection;**
● **Nerve impulses and growth control are controlled by proteins.**

The twenty basic amino-acids are the building-blocks for the proteins in all known species, from bacteria to mankind. The basic alphabet of proteins appears to be at least 3,500 million years old.

NUCLEIC ACIDS
Nucleic acids are a group of naturally occurring polymers of purine and pyrimidine nucleotides. They are derived from two different but closely-related sugars:
● D-ribose, giving **ribose nucleic acid (RNA);**
● and 2-deoxy-D-ribose, giving **deoxyribose nucleic acid (DNA).**

DNA is the molecule of heredity. It is a very long, thread-like macromolecule made up of a large number of deoxyribonucleotides. The purine and pyrimidine bases of DNA carry genetic information, whereas the sugar and phosphate group perform a structural role.

The nucleotide make-up of DNA is somewhat simpler than RNA. The molecular size and weight of DNA specimens are enormous. In primitive

Insulin, the anti-diabetic hormone, has fifty-one amino-acid units and a molecular weight of 6,000. The first protein whose structure was determined, it consists of two polypeptide chains cross-linked.

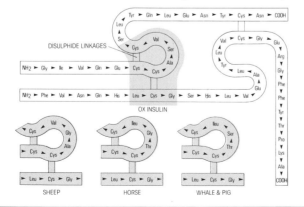

DISULPHIDE LINKAGES

OX INSULIN

SHEEP   HORSE   WHALE & PIG

prokaryotic cells containing a single chromosome, the DNA may be a single molecule with molecular weight greater than 2×10^9. In more complex eukaryotic cells, with several chromosomes and so many different types of DNA, the molecular weights are vast.

Simple bacterial cells, which contain a single molecule of DNA not normally associated with protein, contrast with the DNA of eukaryotic cells which is commonly associated with strongly basic proteins.

The DNA in bacterial cells (which may be as much as one per cent of the cell weight) floats free in the cytoplasm. But eukaryotic cells normally have the DNA concentrated in the cell nucleus.

RNA takes three major forms which co-exist in cells:

● **Messenger RNA (mRNA)** gives the molecular pattern for protein synthesis. There is an mRNA molecule corresponding to each gene or group of genes which is being expressed.

● **Transfer RNA (tRNA)** carries amino-acids in an activated form to the cell-ribosome to be linked with other amino-acids in the peptide bond, in a sequence determined by the mRNA. There is at least one kind of tRNA for each of the twenty amino-acids.

● **Ribosomal RNA (rRNA)** is the major component of cell-ribosomes, where protein synthesis takes place. But its precise role in protein synthesis is unclear.

There are at least three rRNAs, several hundred RNAs and about a hundred tRNA molecules.

DNA is the molecule of heredity. As this computer-generated model indicates, it is a very long macromolecule of thread-like appearance. It is made up of a large number of nucleotides.

Living cells usually have roughly ten times as much RNA as DNA. Where RNA molecules are found in a cell depends largely on what kind of cell it is. In prokaryotic cells, such as bacteria, which do not contain a nucleus but a single-coiled DNA molecule, the RNA is found in the cytoplasm (or material) of the cell. But in more complex cells, such as those in mammalian liver, about eleven per cent of the total RNA is in the nucleus (largely as mRNA), fifteen per cent in the mitochondria (rRNA and tRNA), fifty per cent (largely rRNA) in ribosomes and twenty-four per cent (mainly tRNA) in other organelles. RNA also occurs in all plant viruses and in some bacterial and animal viruses.

POLYSACCHARIDES
Proteins and nucleic acids form a dominant central partnership in the functioning of a cell.

They control its growth, division and metabolism. But the major role of polysaccharides appears to be twofold: storing energy and forming structural or skeletal units in membranes or cell walls.

Some polysaccharides are simple and abundant — starches and celluloses, for example. These differ in structure from proteins and nucleic acids, seldom having side-chains, and this limits their function and activity. Other less well-known, more complex polysaccharides, which occur in the walls of some lower organisms and vertebrates, are much more like proteins and nucleic acids.

Three vital and very common substances are polysaccharides:

● **Starch** is the nutritional reservoir in plants; more than half the carbohydrate we eat is in this form.

● **Cellulose** serves a structural role in plants. It is the most abundant organic compound in the biosphere, containing half of the organic carbon.

● **Chitin** is what the external skeletons of insects and crustacea are made of.

LIPIDS
Lipids are a vaguely-defined group of naturally occurring fatty substances, insoluble or almost insoluble in water, but soluble in organic solvents. They include fatty acids, and when these or their derivatives contain paraffin carbon-chains longer than about four carbon units, they become insoluble in water. They thus form a suspension in water, are excluded from the water structure and coalesce to form distinct units. It is this physical

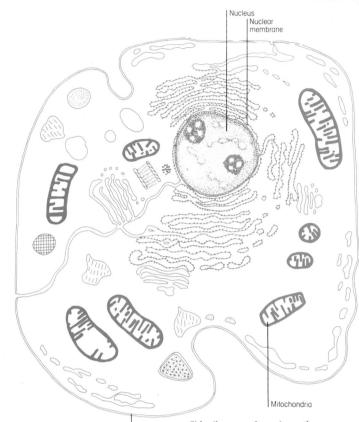

Nucleus
Nuclear membrane

Mitochondria

Plasma membrane

This diagram of a eukaryotic animal cell shows that, unlike the prokaryotic cells of bacteria or blue-green algae, it has a clearly-defined nucleus, bounded by a nuclear membrane.

property which is vital for forming cells.

CELLS

There are two fundamentally different types of cell into which all living systems can be classified:

● **Prokaryotic cells** (*pro*, meaning 'before', and *karyon*, a 'kernel') do not have a nucleus separated by a membrane from the rest of the cell cytoplasm. Blue-green algae are prokayotes, as are bacteria (and actinomycetes).

● **Eukaryotic cells** (*eu* meaning 'well') have a distinct nucleus. All other organisms are eukaryotic.

The difference between these two types of cell is accepted as the deepest division in biology:

● **Prokaryotic cells** have no nucleus, no mitochondria (the part of a cell which divides), no chloroplasts (containing chlorophyll) or any such organelles. They have a single (usually circular) strand of DNA, and the ribosomes (for protein synthesis) are suspended in the rigid, gel-like cytoplasm.

● **Eukaryotic cells** have their contents (organelles) neatly packaged in distinct units,

which appear to float in the cytoplasm. The cell contains a complex nuclear structure bounded by a membrane, which has large porous openings. The DNA in the nucleus forms the multiple complex of chromosomes. The cells contain several hundred mitochondria — about twenty per cent of the cell volume. The whole cell is bounded by a membrane; the nature of this membrane is now well understood in many plants and fungi, but less understood in animal cells. It appears to be made up of protein and lipids.

VIRUSES

Viruses are not self-sustaining organisms; they cannot be considered as living. They are not primitive, but highly sophisticated parasites on cells.

They have been defined (by Fraenkel-Conrat) in this way:

Viruses are 'particles made up of one or several molecules of DNA or RNA, and usually but not necessarily covered by protein, which are able to transmit their nucleic acid from one host cell to another and to use the host's enzyme apparatus to achieve their intercellular replication by superimposing their (genetic) information on that of the host; or occasionally to integrate their genome in reversible manner to that of the host and become cryptic or to transform the character of the host cell'.

Viruses have often been thought of as at the threshold of life. But they are apparently not the product of chemical evolution from simpler structures. They are either cells gone wrong or a degenerate product of a higher form of life.

Chemical Evolution Theory

E arly theories of the abiotic synthesis of organic chemicals and of the origin of the Solar System by the condensation of a primeval dust cloud led to one common concept. This concerned what the primeval atmosphere was like that existed on Earth at the time it was formed.

The fundamental assumption about this hypothetical atmosphere was that it was oxygen-free (reducing) and contained hydrogen, in direct contrast to the oxygen-containing atmosphere that we know today.

With the development of new analytical techniques for separating and identifying small amounts of organic compounds, an American chemist named Stanley Miller was able in 1953 to try to simulate these processes. He examined the chemical effects of adding different energy-sources to model reducing atmospheres of the types postulated by Oparin and Haldane. The results of these experiments, and many subsequent related ones, are well known. It is clear that from such synthetic reducing or neutral atmospheres it is possible by supplying energy (electric, impact, radiation, heat or whatever) to produce some of the basic organic molecules which are formally required for the production of biomolecules — the 'chemicals of life'.

These molecules include a number of amino-acids, nucleic-acid bases, nucleosides and nucleotides, carbohydrates and other substances. The success of the early abiotic-synthesis experiments has encouraged the escalating interest in the origin of organic molecules on the primitive Earth — and thus of life itself. These experiments have produced a general theory to explain the origin of the living system. It is called the 'Chemical Evolution Theory', and is often quoted as if it were proven.

This theory may be briefly summarized thus:

● **The primitive Earth had a reducing or at least a neutral atmosphere.**

● **There were sources of high energy (ultra-violet, heat, electric, impact and so on) available to the system.**

● **Given these conditions it was inevitable that the sorts of organic chemicals mentioned above — especially**

Micro-organisms are plentiful in these hot springs in Yellowstone National Park, Wyoming. According to chemical evolution theory, these hot, chemically-charged environments would have been ideal sites for life to begin.

amino-acids, nucleosides and nucleotides — would be produced.

● Since there was no life on the planet to metabolize these organic materials (as would be the case today), such materials would accumulate in the Earth's environment, forming what has become known as the 'primitive soup'. Estimates of the strength of this 'soup' vary from virtually nothing to as much as one gram per litre.

The environment in which all this is supposed to have occurred has been discussed lately, and there is much conflict of opinion. One hypothesis can be designated the **marine hypothesis**, and the other the **volcanic hypothesis**. Both hypotheses were first explored by Oparin in his first paper on the origin of life, but the marine hypothesis gained its greatest impetus from Haldane's concept of a 'hot, dilute soup'.

These hypotheses suggest different energy-sources as the power for the evolutionary processes that later gave rise to life. The marine hypothesis postulates solar energy, as the Sun acts on water, whereas the volcanic hypothesis draws mainly on terrestrial energy, from volcanoes. The marine hypothesis frequently postulates that the first forms of life must have been 'autotrophic' (self-nourishing), whereas the volcanic hypothesis insists that they must have been 'heterotrophic' (drawing nourishment from outside themselves). The organic precursors of life according to the marine hypothesis can be relatively small molecules that exist in very dilute solutions; the volcanic hypothesis invokes more concentrated solutions, the presence of unmixable polymers, and the prebiological synthesis of suitable food materials.

● The next stage of the general theory requires a long period of time, during which **these organic molecules interact through chance collisions**. The frequency of these collisions increases as the aqueous solution becomes more concentrated.

An alternative mechanism would involve the evaporation of the 'soup' to produce small pools or even dried-up residues with very high concentrations of organic matter. These dry residues could produce polymeric materials under more favourable conditions than might exist in an aqueous environment, where hydrolysis (decomposition by reaction to water) will always tend to be the major chemical reaction. But this suggestion has a disadvantage: the processes would have to take place under an essentially water-free condition.

Overleaf
For the chemicals of life to have come into existence by abiotic synthesis, large sources of energy would have had to have been available. Lightning is one possible means by which such energy could have been unleashed.

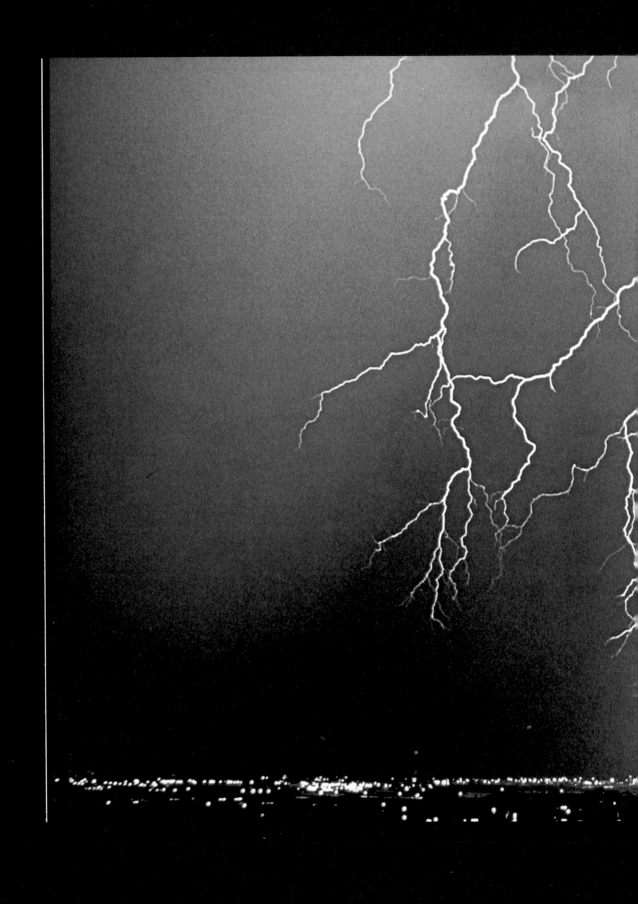

● The next phase envisages **these organic molecules further interacting to form large complex bio-molecules of high molecular weight**. The process can be summarized as follows:

Amino-acids ⟶ proteins (molecular weight approximately 50,000);
Nucleotides ⟶ nucleic acids (molecular weight approximately 1 million).

Abiotic polypeptide-like materials have been produced by many workers by subjecting amino-acids to a variety of conditions, including heat-treatment and chemical-condensation reactions. Fox, in 1976, reported experiments in which amino-acids reacted to form possible models for so-called 'pre-proteins', known as 'proteinoids'.

There have been attempts to progress towards an abiotic chemical synthesis of the whole or part of a nucleic-acid molecule. Some limited progress has been made in this area of chemical-evolutionary experimentation, with the synthesis of polymers with small chain lengths.

● In considering these various prebiological experiments, it is often suggested that **Miller's experiments support the marine hypothesis, whereas those of Fox support the volcanic hypothesis.** But Sylvester-Bradley pointed out that this is not quite true, since Miller-like experiments can readily occur in a volcanic environment.

It is also sometimes claimed that one hypothesis involved a 'cold and wet' environment, whereas the other was 'hot and dry'. But this too is very far from the truth, for the volcanic hypothesis involves as an essential step the quenching of hot, organic material with liquid water. Proteinoids have been produced only when conditions in the laboratory experiments oscillated between dry and wet, the material alternately dehydrating and rehydrating.

● The next and proposed final stage in Chemical Evolution is a fundamental step. **In some as-yet-unexplained way the nucleoproteins, proteins, other essential components, and the membrane material form a complete living cell unit apparently with few intermediate forms (proteinoids apart).**

The problem here is not that of synthesizing the aggregates that are considerd to be possible progenitors of

The apparatus used by Stanley Miller in 1953, in which amino-acids were first synthesized from a mixture of gases.

COMMON AMINO-ACIDS

These are the twenty amino-acids most commonly found in proteins, together with their common abbreviations

Amino-acid	Three-letter abbreviation	One-letter symbol
Alanine	Ala	A
Arginine	Arg	R
Asparagine	Asn	N
Aspartic acid	Asp	D
Asparagine or aspartic acid	Asx	B
Cysteine	Cys	C
Glutamine	Gln	Q
Glutamic acid	Glu	E
Glutamine or glutamic acid	Glx	Z
Glycine	Gly	G
Histidine	His	H
Isoleucine	Ile	I
Leucine	Leu	L
Lysine	Lys	K
Methionine	Met	M
Phenylalanine	Phe	F
Proline	Pro	P
Serine	Ser	S
Threonine	Thr	T
Tryptophan	Trp	W
Tyrosine	Tyr	Y
Valine	Val	V

cells. About two hundred co-acervates (the droplets that form these aggregates) have to be synthesized for a living cell and in itself that is not impossible. But for such syntheses to be a logical step in the abiotic synthesis of life involves 'raising a cell by its own bootstraps'. Proteins cannot be synthesized without DNA, but you cannot make DNA without enzymes, which are proteins. It is a kind of chicken-and-egg situation.

Enzymes are complex protein molecules and it is impossible to imagine that such a specific compound could have been produced by chance, even in a period of up to 300 million years. Professor Quastler has calculated that the odds against producing a specific complex molecule on Earth are 1 in 10^{301} (10 followed by 301 zeros), which is very near to impossible. Other calculations have been made to estimate the chance probability of a DNA molecule being produced somewhere in the Universe. If one assumes that there could be 10^{20} planets in the Universe where life may exist or have existed, then the odds of a complex DNA molecule being formed by chance are 1 in 10^{415}, and these odds lengthen to an astonishing 1 in 10^{600} if a longer strand of DNA is postulated.

It is very hard to imagine even the complex molecular building blocks of life (proteins, enzymes and DNA) being formed by chance. And the random abiogenic origin of a simple living cell is approaching the impossible.

Most scientists now accept this conclusion and in recent years an increasing amount of research work is being carried out into more complicated interrelated chemical processes where patterns of interaction between different molecules on different surfaces and in different conditions could possibly interlock two or more systems together.

Experimenters, then, have repeatedly been able to demonstrate the formation of interesting organic chemicals under synthetic laboratory conditions. But these are only laboratory experiments; they do not really tell us for sure what happened on the primitive Earth.

The PreCambrian Rocks and Early Life

T he last 600 million years of the Earth's history have been well documented by palaeontologists and other investigators. There is now an improved understanding of the nature and evolution of various biological systems. As part of this, there has in recent years been a markedly-increased interest in PreCambrian rocks and their organic contents.

The PreCambrian Era spans the time between the origin of the Earth and the beginning of the Cambrian Era (see *The Geological Column*). It lasted from about 4,500 to about 600 million years ago, and accounts for approximately seven-eighths of geological time. This period includes at least eighty per cent of the time required for the presumed origin, development and evolution of living systems — probably the most important part. So what can these rocks tell us?

The studies of PreCambrian organisms and organic matter were pioneered in 1883 by C. D. Walcott, then the acknowledged leader in the early search for evidence of biological fossils in the PreCambrian. He it was who first suggested that PreCambrian laminated stromatolites were probably of algal origin. But Walcott's work aroused scepticism among contemporary palaeontologists, as did the later work of Gruner in 1923–25, who claimed to have discovered filamentous micro-organisms in PreCambrian rocks. During the last three decades, however, detailed examinations of PreCambrian rocks have conclusively demonstrated the presence of micro-organisms and fila-mentous remains. All the available data shows that Walcott and Gruner were right. Organic micro-organisms do exist within the cherts and stromatolites of the PreCambrian. Walcott's and Gruner's general theses now form the basis of many current PreCambrian investigations.

All this helps a great deal in investigating what happened during the early stages of the Earth's history. Clearly, one of the main hopes of reaching precise conclusions about the validity of Chemical Evolution and alternative theories lies in examining materials that have existed more or less unchanged since almost the beginning of the Planet. Hopefully these materials might have retained in their structures a record of the chemical and physical happenings of the time. Such materials include ancient sediments which are found through the world in various locations.

PRECAMBRIAN ORGANIC MATTER

This PreCambrian volcanic rock is from the Archaean Age, about 2,700 million years ago. It is the Abitibi Greenstone belt in the Quebec Province of Canada. The green colours are produced by copper minerals; the browns and yellows by iron minerals.

Various kinds of once-living systems have been identified in the PreCambrian. They include:

- **structured organic micro-organisms and filaments;**
- **unstructured or partly-structured organic matter**, which can very often be chemically related to materials of known biological origin;

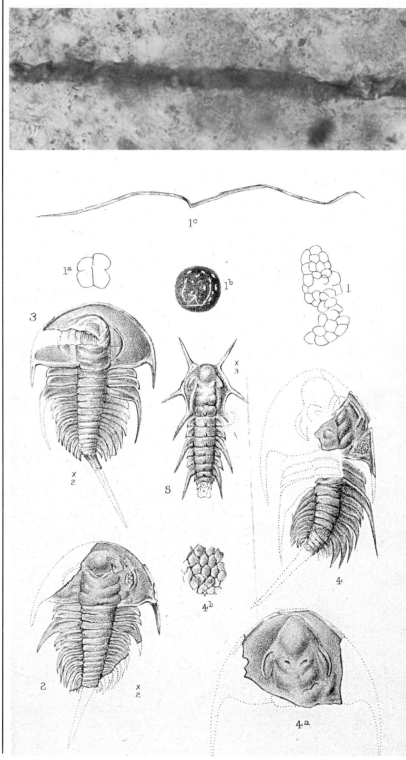

These microfossils were recovered from PreCambrian sediments in the Torridon area of Western Scotland. They show filamentous and structured micro-organisms. Their age is 850–900 million years.

● **extractable organic compounds** with chemical structures characteristic of biologically-produced materials.

Although most PreCambrian rocks were heated up and cooled down to varying degrees, the organic matter they contain is often still sufficiently preserved to allow meaningful geochemical and palynological examination. Well-preserved micro-organisms have been described from many areas, representing all PreCambrian periods. They are considered to be remains from bacteria, algae and possibly aquatic fungi.

Geochemical and palynological studies have been conducted in the Early PreCambrian (often called the Archaean Era — older than 2,500 million years). There have been four main areas of study:

● **Stromatolites;**
● **Banded-Iron Formations;**
● **Rocks for chemical traces of life;**
● **Micro-organisms.**

The cherts in which these micro-organisms were found have been radiometrically dated accurately at 3,400 million years old. The cherts are in the Onverwacht series in Swaziland, Southern Africa. The micro-organism is 15–20 microns in diameter.

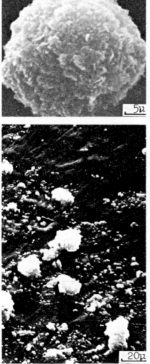

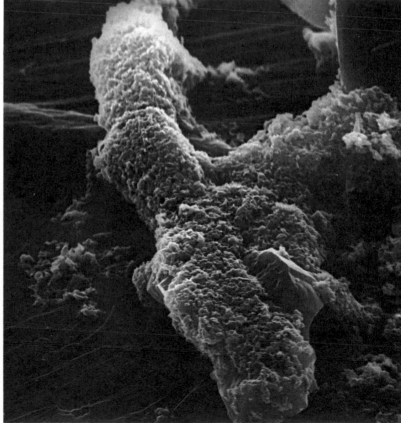

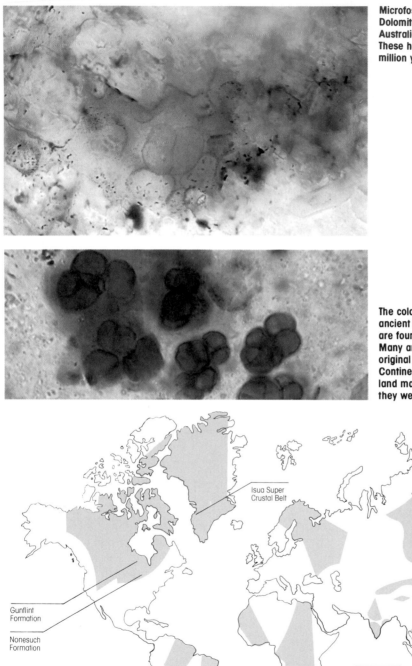

Microfossils from the Amelia Dolomite, McArthur river area in Australia's Northern Territory. These have been dated at 1,700 million years old.

The coloured areas show where ancient PreCambrian platforms are found in the world today. Many are a long way from their original positions, because Continental Drift has moved the land masses considerably since they were first deposited.

Isua Super Crustal Belt

Gunflint Formation

Nonesuch Formation

McArthur Basin

North Pole Pilbara Block

Fig-tree & Onverwacht Formation

Bitter Springs Formation

Stromatolite domes are formed by the activity of micro-organisms. These domes are in Peary Land, North Greenland. They are about 900 million years old.

Archaean Stromatolites (up to 3,500 million years ago)

The word 'stromatolite' has a restricted meaning. It is applied to organo-sedimentary structures predominantly accreted by sediments that have precipitated as a result of the growth and metabolic activities of prokaryotic micro-organisms on the ocean-floor.

As early as 1858, W. E. Logan suggested that stromatolite structures were evidence for PreCambrian life, and since then many studies and correlations have been carried out on these structures. Various authors consider that stromatolites offer the most compelling evidence for the existence of life, since similar modern structures are well known and have been studied in detail (at Shark Bay, for example, in Western Australia). Stromatolites are found from the Archaean Era right through to present-day deposits. The oldest-reported compelling evidence for life on Earth comes from stromatolites about 3,400–3,500 million years old from the Warrawoona Group in the Pilbara Block of 'North-Pole' Western Australia.

A chert unit is extensively found near the top of the 3,400 million-year-old Warrawoona Group. This chert formed by silicification of a carbonate-evaporite sequence deposited in shallow water just above and below low-water mark. It is characterized by internally-laminated conical moulds.

Studies of the structure and internal organization of these moulds suggest that they are conical stromatolites. They indicate that there were microscopic life-forms on the ocean-floor 3,400–3,500 million years ago, but it is not known quite what they were.

Banded-Iron Formations (up to 3,800 million years old)
Siliceous Banded-Iron Formations (BIFs) are widely distributed in PreCambrian rocks. BIFs include a variety of rock-types, but all contain iron and are often cherty. Micro-organisms occur in some BIFs; they are found among time-regular bands of layered cherts containing many layers rich in iron oxide, giving iron-rich and iron-poor microbands. These were caused as nutrients from algal blooms of ferrous iron in solution welled up from time to time. Layered BIF cherts occur in some Archaean rocks, but they are found on a much larger scale in slightly younger rocks — the Hammerley Group in Western Australia, for example.

The oldest reported BIFs have been identified in metasediments in West Greenland which are about 3,800 million years old. These include what is undoubtedly BIF some hundreds of feet thick and cropping out over tens of miles. McGregor suggested in 1973 that the rocks of the Godthaab region and of the Isua mantled-gneiss dome in West Greenland were derived by deformation of granites which contained inclusions of older rocks. Among these are rocks that are probably relics of BIFs. If so, they are among the oldest sedimentary rocks in the world, and can scarcely be much younger than 4,000 million years old.

The presence of BIFs in Archaean rocks is taken as an indication that micro-organisms were active during the early PreCambrian. This confirms early plant evolution. It suggests that not only life itself but also a low level of oxygen-producing photosynthesis was already in existence more than 3,800 million years ago, when the oldest known BIFs were being deposited in South-west Greenland.

Micro-organisms, filaments and organic chemicals
Two groups of rocks are especially interesting here:
●   **The Swaziland Supergroup of Southern Africa** (3,100–3,400 million years old) consists of a complex of folded sediments. It has three major groups: the Moodies Group at the top of the sequence, the Fig-Tree in the middle

Stromatolites in a dolomite unit, in the Yellowknife Supergroup, Slave Province, Canada, dated at 2,650 million years old. Such units can be up to 130 feet thick and have been traced for as much as four miles.

and the Onverwacht at the base. The Onverwacht Middle-Marker Bed, which occurs at the base of the Upper Onverwacht, has been radiometrically dated at 3,355 million years old.

The organic chemicals and residue extracted from Onverwacht rocks have been analysed geochemically. Micro-organisms with various morphologies and filamentous structures have been identified within them. Several types of spheroids, filaments and even colonial structures have

The Pilbara Block of the Warrawoona Group in the North Pole region of Western Australia contains the most compelling evidence for very ancient life. It is dated to about 3,556 million years ago.

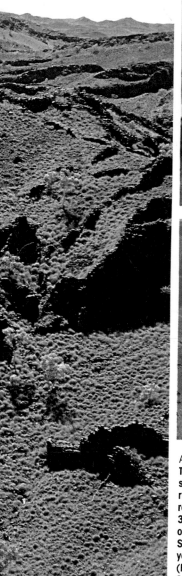

Above right
The Barberton Mountain land showing the Early PreCambrian rocks in Swaziland. The lower rocks (Onverwacht Series) are 3,200 to 3,540 million years old; the middle rocks (Fig Tree Series), 3,000 to 3,200 million years old; the upper rocks (Moodies Group), 2,800 to 3,000.

Top three photos
Micro-organisms of fossil blue-green algae, from the Kromberg Formation of the Onverwacht Series cherts in Swaziland, dated at about 3,200 million years old.

Right
These micro-organisms from the Theespruit Formation are probably the oldest to have been identified in the Onverwacht Series (about 3,550 million years old).

been described in the Onverwacht Group. Statistically, morphologically and chemically, the spheroids in these sediments appear to have been produced by living materials.

● **The Isua and Godthaab Metasediments of South-west Greenland** are about 3,800 million years old. The geological history of the Isua supercrustal belt has been studied and radiometrically dated. The Archaean Craton of South-west Greenland contains some of the oldest known non-igneous rocks. It includes the Amitsoq gneisses of the Godthaab

A view looking towards the Godthaab region of Western Greenland shows the world's oldest known rocks. The foreground contains 2,800 million-year-old granulite gneisses. A fault runs down the fjord. The mountains in the background are mainly 3,700 million years old — Amitsoq gneisses representing the oldest part of the continental crust so far to have been reliably dated.

WAS THERE LIFE 3,700–3,900 MILLION YEARS AGO?

In recent years several reports have suggested the possible existence of life in the Isua, West Greenland about 3,800 million years ago.

Published scientific data do not allow a decisive choice to be made about the sedimentary organic matter in these most ancient rocks. Was the organic matter biologically produced or abiogenically synthesized? Or could it have been derived metamorphically from inorganic carbonate minerals?

The organic matter in the Isua rocks is complicated to interpret. To say definitely that living systems were (or were not) extant in Isua time is to go beyond direct scientific evidence. But it is the subjective opinion of a number of scientists that life may well have existed in Isua times, some 3,800 million years ago.

Current evidence shows that living systems must have existed earlier than 3,500 million years ago, but how much earlier we can only speculate at present.

Some interpretations of the nature of the Isua rocks suggest that photosynthesis had evolved to a high degree. Even if we do not yet fully accept these Isua interpretations, there is evidence that photosynthetic organisms existed in rocks at Onverwacht (around 3,300 million years old) and in the Warrawoona Group (around 3,400 million years old).

This is a puzzling fact, because the change from an oxygen-poor to an oxygen-rich atmosphere is not considered to have occurred until 2,000 million years ago (although some scientists, including myself, consider that an oxygen-rich atmosphere was present much earlier than often predicted). If photosynthesis was going on so very long ago, this implies very rapid chemical evolution, since the number of simpler chemical processes are thought to have evolved into the complex process of photosynthesis. It also poses another question. If photosynthesis was a 'big leap forward' in biochemical efficiency, why did many organisms that do not use oxygen persist so long?

This puzzle is more complicated if we assume that the Earth's early atmosphere was oxygen-rich during the Early PreCambrian. Dr Ken Towe (Smithsonian Institute, Washington) has commented: 'The study of PreCambrian Earth history presents an enigma. On one hand, it is widely conceded that the primitive Earth was initially devoid of molecular oxygen and that life originated in such an environment. On the other hand, many PreCambrian rocks, including the oldest known sediments, contain primary oxidized iron minerals which indicate some source of free oxygen at the time of their precipitation and deposition. Where did this free oxygen come from?'

A number of sources have been proposed and the general scientific opinion is that the free oxygen came from the dissociation of water molecules in the early atmosphere by ultra-violet light rather than via photosynthetic bacteria. (See *Biosphere, Atmosphere and Hydrosphere*.)

In the postulated reducing atmosphere (without oxygen), it is proposed that the earliest life-forms (bacteria) functioned by using methane as their energy source. The suggested early life-forms, *Archaebacteria* or *Methanogens*, probably soon disappeared. This was because first fermentation and later the appearance of the chlorophyll molecule and photosynthesis made it possible to use energy from the Sun directly in biochemical processes.

But what if the primitive atmosphere was not reducing but contained oxygen? This causes major questions to be asked about chemical evolution theory and the way in which life originated on Earth. Many alternative hypotheses have been proposed. Some suggest ways in which ammonia could have been generated and utilized. Others consider that methane rather than ammonia was the main constituent.

These questions, puzzles and alternative hypotheses show that the scientific view of the origin of life in the Early PreCambrian is still highly speculative. There is a long way to go before we can know for sure what happened.

region and the markedly low-grade Isua belt of metasediments and metavolcanic rocks, which appear to have
remained at substantially lower temperatures and retain
some primary structures.

In addition to the BIF in the metasediments, there are also
marble (metamorphosed limestones) and graphitic mica
schist (metamorphosed organic-rich shale). Cell-like material
has been detected in the cherty layers, and this was analysed
by Pflug in 1979 using a Raman laser-molecular microprobe.
He interpreted it as consisting of biological materials.

The fossils occur as individual unicells, filaments or cell
colonies. Cells and cell families are usually surrounded by
multilaminate sheaths that show a characteristic laminar
structure. They all apparently belong to the same kind of
organism, named *Isuasphaera*. Pflug suggests they are
related to recent yeast cells. But Bridgewater and colleagues
reviewed these reports in 1981 and concluded that such an
interpretation is inconsistent with what is known about these
rocks and with the structure of the materials themselves.
They consider the microstructures present in the Isua
samples to be indistinguishable from inorganic limonite-
stained fluid inclusions, originating since the rocks were
deposited, and as such should not be regarded as evidence of
early Archaean life-forms.

A similar fate has befallen another suggestion — that
hydrocarbons extracted from Isua rocks are remnants of
organisms that existed about 3,800 million years ago. Recent
work by Nagy (1981) shows that such an interpretation is
inconsistent with the high-temperature history of the rocks,
and that this material is really of much more recent origin.

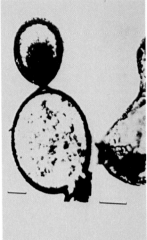

Are these the oldest known fossils? Microstructures Isuasphaera isua Pflug 1978, **they were identified in the Isua Series of South-west Greenland and dated at around 3,800 million years old. Their interpretation has given rise to extensive discussion, but it is not possible so far to be conclusive as to whether these are fossils of once-living organisms.**

Where is the Evidence for Chemical Evolution?

The Chemical Evolution Theory requires that an oxygen-free atmosphere existed for a considerable period of time on the early Earth. Geological evidence from the early PreCambrian, however, suggests that such primitive and secondary atmospheres did not exist for any appreciable length of time.

Supporting evidence for this comes from studies of the ultra-violet photolysis of methane to give polymeric materials. These studies suggest that under primitive-Earth conditions the temperature around the Earth might have been so high that methane would have disappeared. It would have broken

The Grand Prismatic Spring in Yellowstone National Park, USA, provides the sort of environment where micro-organisms may well have existed in Early PreCambrian times.

down into high-molecular-mass carbon polymer — deposited on the Earth's surface, and hydrogen — escaping instantaneously into space. So an oxygen-free atmosphere on primitive Earth, if it existed, would probably have broken down in too short a time for a living system or chemicals of life to have formed in it. These observations led Shimizu to conclude, in 1976, that the previous optimistic conclusions as to primitive Earth's atmosphere should be cautiously re-evaluated.

And then what of the 'primitive soup' required for Chemical Evolution? If such an environment ever existed on Planet Earth for any appreciable time, it would require relatively large quantities of nitrogen-containing organic compounds (amino-acids, nucleic acid bases and so on). It is likely that such nitrogen-rich soups would have given significant quantities of 'nitrogenous cokes', trapped in various PreCambrian sediments. (The formation of such 'cokes' is the normal result obtained by heating organic matter rich in nitrogenous substances.) No such nitrogen-rich materials have yet been found in early PreCambrian rocks on this planet. In fact the opposite seems to be true: the nitrogen content of early PreCambrian organic matter is relatively low (less than 0.15%).

From this we can be reasonably certain that:

● **there never was any substantial amount of 'primitive soup' on Earth when ancient PreCambrian sediments were formed;**

● **if such a 'soup' ever existed it was only for a brief period of time.**

Subtract from the basic concept of the Chemical Evolution Theory the ideas of substantial amounts of 'primitive soup' and a long period of time, and there is very little left.

And yet there is good evidence in the rocks to indicate that throughout the early PreCambrian living systems were present. They were there at the time when the rocks were deposited, and they were probably photosynthesizing and undergoing biochemical reactions similar to those of current living systems. The oldest preserved sediments probably formed about 3,900 million years ago, before the major metamorphic events dated at 3,750 million years ago. The BIFs in the Isua and Godthaab metasediments and the stromatolites and micro-organisms in the Pilbara Block of Western Australia suggest that living systems were probably active about 3,800–4,000 million years ago. Before this time the Earth's surface is considered to have been at too high a

These artist's impressions give a geological understanding of how the early Earth might have looked at different stages. The top picture shows the Earth's semi-molten surface some 4,500 million years ago. By 4,000 million years ago (middle picture) a solid crust has formed; it is shown pockmarked with meteorite craters. In the bottom picture we see the shores of an early ocean, about 3,800 million years ago.

Life is thought to have begun on Planet Earth in the period about 3,800 million years ago. The oldest known Earth rocks, in Greenland, were probably formed about 3,750 million years ago.

temperature (more than 600°C) to support life, or for that matter to allow the stable existence of complex biomolecules such as proteins and nucleic acids. This leaves ever-decreasing amounts of time for conventional chemical-evolution processes to have occurred. It is a time-scale very different from that normally suggested for chemical-evolution models.

All the chemical-evolution hypotheses are speculative. It is quite possible that they have no bearing on the origin of life on Earth. Life probably originated on Earth at least 3,800 million years ago, and the evidence for its origin is no longer available. The chemical-evolution models are not yet proved and we cannot find PreCambrian rocks old enough to test their ideas. None of the vast quantities of early PreCambrian rocks contain anything that can be positively recognized as pre-biological organic matter.

Since there appears to be no clear evidence on Earth to support current theories of Chemical Evolution, where else can we look to explain the existence of early life? The organic constituents of meteorites and of interstellar molecules may provide useful clues.

Meteorites

The Solar System contains many solid bodies. Their sizes vary from the planets through the smaller but still-large satellite moons and asteroids to microscopically small dust particles. This solid matter follows various orbits round the Sun, and the smaller particles especially are consistently dragged into the gravitational field of their fellow-bodies, including the Earth. It is estimated that each year the Earth receives about 100,000 tonnes of material from space. Much of this solid matter is rapidly destroyed after entering the Earth's upper atmosphere, by frictional heat and combustion.

We call such material **meteors** or **meteoroids**. Their incandescent deaths, marked by brilliant streaks of light, are the 'shooting stars' of the night sky. Estimates of the total number of meteorite falls have varied, but a recent well-supported figure is about 3,500 annually.

Only a few of the larger meteors (or their fragments) survive the violent passage through the atmosphere and reach the Earth's surface, and even then about three-quarters fall into the oceans. These survivors are called **meteorites**. Their outermost surfaces are heated to high temperatures during descent, but the low conductivity of their material to heat frequently means that, especially in reasonably large specimens, their inner parts have remained unaltered.

Other visitors are the **comets**. These are considered to be either solid 'dirty snowballs' or, less likely, swarms of interstellar dust and ice-particles held together by their mutual gravitational attraction. Comets come from the cold outer Solar System. They are composed mainly of chemical compounds that are solid in space but would exist as gases on the Earth. Most of the comet is composed of water ice, frozen ammonia, methane and carbon dioxide; together in their frozen state they give the appearance of 'snow'. When comets pass close to the Sun, they are bombarded by the solar wind (atomic particles) and also heated. This changes the frozen-solid gases into vapours (which sometimes explode), and causes the comet to break up into microparticles or dust. In this way a meteor stream is born.

A meteorite fall may be seen as a bright fireball with long, incandescent trails of debris. It may be heard as a thundering, whistling or cracking sound, sometimes accompanied by loud supersonic bangs. Meteorite sites, of which over 2,000 are now known, are scattered at random on

This artist's impression of a meteorite fall and fireball conveys the spectacular effect of these rocks from the depths of space entering Earth's atmosphere. Their arrival is irregular and unpredictable.

Comet Ikeya-Seki **entering the Earth's atmosphere after its journey through the Solar System from the cold outer reaches of our planetary system.**

the Earth's surface. Evidence of meteorite impacts on the Earth's surface is rare, because normal geological processes gradually remove the evidence. However, large meteorites (often weighing more than a hundred tonnes) occasionally have hit the Earth, and on impact produce explosion craters. World-wide investigations suggest that up to sixty meteorite craters have been identified on Earth. The best-known and most frequently visited and photographed meteorite-explosion crater is 'Meteor Crater' in northern Arizona, USA. This crater is estimated to be 20,000 years old. It is 1.2 kilometres across, and 170 metres deep from the top of the raised rim to its floor.

Meteorite falls are rare, and of course they are unpredictable as to time and place. This makes planned observations extremely difficult. But in 1959 a bright fire-ball was photographed by synchronized cameras designed for the study of meteorites. Czechoslovakian astronomers managed to obtain precise measurements of the speed and path of the Pribram meteorite that fell on 7 April 1959. Two other meteorites — the Lost City (Oklahoma) fall of 8 January 1970 and the Innisfree (Alberta) meteorite fall of 5 February 1977 — have provided precise information on their trajectory. From these photographs, it has been worked out that the meteorites were travelling in elliptical orbits whose farthest points were between Mars and Jupiter. Since about this time, meteorites have been considered as asteroids or fragments of asteroids that have been captured by the gravitational field of the Earth. But these three recent observations are the first instances of positive verification.

The Leonid meteor shower happens regularly every year as the Earth passes through a band of small particles orbiting the Sun. This photograph was taken on 17 November 1966.

CARBONACEOUS CHONDRITES

Currently meteorites are divided into three broad classes, which differ in their metal silicate content: irons, stones and stony-irons. The most interesting group of stony meteorites are the **carbonaceous chondrites**, which are of special importance because they contain organic matter (up to 5%) and associated hydrated minerals.

About forty carbonaceous chondrites have been found, and these rocks are currently the only extra-terrestrial material containing organic matter available for scientists to study. These chondrites are also the oldest known matter. Radiometric dating has repeatedly shown them to be 4,500–4,700 million years old, which makes them at least 600 million years older than the oldest dated rocks so far found on Earth. Meteorites thus contain an ancient record whose counterpart may have been destroyed on the Earth's surface by geological events.

The first known carbonaceous-chondrite fall occurred at Alais, France on 15 March 1806. This chondrite was studied

These three extra-terrestrial rocks all broke on impact with the Earth, but a good proportion of their matter was recovered and sometimes re-assembled.

Top left is the Murchison carbonaceous chondrite, which fell at about 11 a.m. on 28 September 1969 near Murchison, Victoria, Australia. About 225 kilograms of its matter was recovered over an area of fourteen square miles (the largest fragment was 2.5 kilograms). This meteorite contains about two per cent organic matter.

Top right is the Barwell meteorite which fell on Christmas Eve 1965 on the common land leading to the centre of Barwell in Leicestershire, England. The twenty-six fragments, collected over about three-quarters of a square mile, weighed together at least forty-four kilograms. Several fitted together perfectly along cleavage surfaces.

The third meteorite is a carbonaceous chondrite again, the Orgueil carbonaceous chondrite which fell in France in 1864. About ten kilograms of it was recovered. This meteorite contains about seven per cent organic matter.

by the famous Swedish chemist Berzelius, who later described an aqueous extract and pyrolysate of the organic material it contained. He asked:

'Does this carbonaceous stone contain humus or other organic substances?'

'Does this possibility give a hint concerning the presence of organic structure in other planetary bodies?'

Similar questions are still asked today.

Since this first study, numerous chemical analyses of organic compounds in meteorites have been carried out. In 1961 Claus and Nagy reported the presence of small spherical objects in samples of the Orgueil and Ivuna meteorites and concluded that these organized elements may be microfossils indigenous to the meteorite. These opinions aroused great opposition because the chemical composition of the meteorites did not resemble that of typical microfossil-containing terrestrial sediments.

During the 1960s, additional claims for finds of structured bodies were made by various groups in studies with crushed fragments, thin sections and acid-resistant extracts of the Orgueil, Murray, Alais, Tonk, Ivuna and Mighei carbonaceous chondrites. (The ordinary stony meteorites, Bruderheim and Holbrook, lacked objects.) Some of the microfossil objects had simple structures, while others were more complex. Finch and Anders, in 1963, claimed that some of these objects were recent pollen- and spore-contamination, and also that the structured bodies are most likely to be mineral grains and hydrocarbon droplets as well as contamination.

Brooks and Muir examined the structured objects in the Orgueil and Murray meteorites by scanning electron microscopy in 1971 and found that the organic box-like structures have five or six sides. The basic symmetry of all the structures is six-sided and those structures with a different number of sides, whether it be five or seven, are either imperfect or probably damaged during preparation. The structures are hollow and contain mineral matter, and it is postulated that they were produced by organic matter condensing round a hexagonal mineral grain. Similar structures are present in the Allende meteorite. These structures show no similarities or relationship to any known biological organisms on Earth.

More recently Dr Hans Pflug has examined the organic structures present in the Murchison carbonaceous chondrite, which fell in Victoria, Australia on 28 September 1969.

He reported that the meteorite contained structures similar to those found during the 1960s in the Orgeuil and Ivuna. These structures have led to much discussion and, although Fred Hoyle has interpreted them as 'bacteria' and 'viruses' in his recent book *The Intelligent Universe*, they are not generally accepted by scientists as proving the existence of extra-terrestrial life. Probably the best comment on the structures in the meteorite was made by Hans Pflug himself when he first saw them under his microscope: 'You must make up your own mind. I can only show you the pictures.'

Extractable organic compounds in carbonaceous chondrites have been studied and show that the insoluble organic matter and perhaps a small portion of the solvent-extractable organic material are indigenous to meteorites. Several reports suggest that there are amino-acids in carbonaceous chondrites. This is very important for the chemical-evolution theory of carbon compounds. Results indicate that the amino-acids from the surface of meteorites are largely

These structured 'organic objects' are residues from the Murray and Orgueil carbonaceous chondrites. Note that each is a six-sided structure.

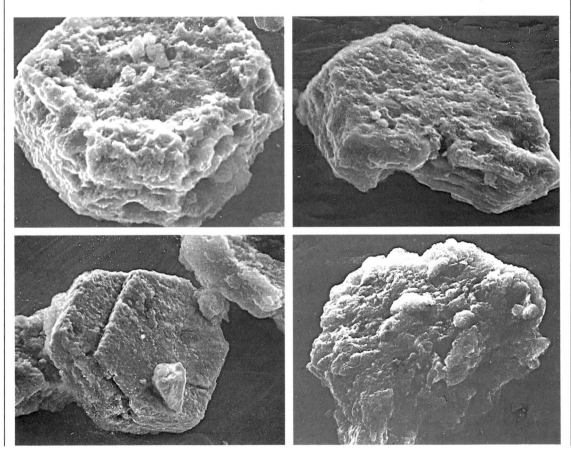

terrestrial contamination, but that the amino-acids from the interior are probably indigenous to the meteorite. Comparing amino-acids in different meteorites and in lunar samples suggests a common mode of origin. These data are readily used to support the view that meteorite organic compounds are the products of an exclusive chemical synthesis from simple precursors. Of course we do not know how this synthesis took place on meteorites, any more than on primitive Earth. But the presence of ubiquitous molecules among the meteorite organic compounds suggests at least some possible analogy between the processes.

Examining the organic matter in the Allende meteorite, it was found that the organic material associated with the fusion crusts was probably derived from formaldehyde, together with small quantities of hydrocarbons and amino-acids. How did this formaldehyde get there? It could represent part of the initial make-up of the meteorite, or it could be more recent. Alternatively, formaldehyde and formaldehyde polymers are known to exist in interstellar space; possibly it may have been absorbed onto the meteorite surface. Another possibility is that the formaldehyde was formed by abiogenic synthesis on the surface of the meteorite. Or it could have been absorbed during its fall through the Earth's atmosphere, though this is considered unlikely.

If we accept that carbonaceous chondrites contain some of the important chemical precursors of life (as for example amino-acids, formaldehyde, hydrogen cyanide), then could they provide a pathway by which the surface of the Earth received significant amounts of these substances? It is estimated that heavy meteorite bombardment onto the early Earth's surface took place 3,800–4,000 million years ago. Such processes are likely to have contributed large quantities of organic compounds to the Earth's surface before 3,800 million years ago. Rough estimates can be made of the amounts of amino-acids, formaldehyde and organic polymers that could have reached the Earth from meteorites. It seems that approximately 0.2×10^{14} g of amino-acids, 10^{13} g of formaldehyde and 10^{18} g of organic polymers could have reached the Earth's surface between 3,800 and 4,000 million years ago.

What Happened to the Dinosaurs?

O ccasionally in the history of life on Earth, whole species have rapidly died out. Why? Could extra-terrestrial influences have been the cause? What may have caused the extinction of the dinosaurs, for example?

Suggestions have been legion: because of the catastrophic impact of meteorites, because the Sun has a previously-unknown dwarf companion which disturbs the Solar System's cometary cloud, or because the Solar System as a whole oscillates above and below the plane of the galaxy. Some have put forward the idea that during the past 250 million years there has been a statistical regularity in the timing of the disappearance of large numbers of species from the surface of the Earth.

Traditional Darwinian theory says that organisms evolve and become extinct primarily as a result of genetic competition, with changes in the physical environment being of subordinate importance. Yet it has become increasingly apparent in the last few years that changes in forms of life through time are characterized by long periods of relative stability punctuated by geologically brief episodes of mass extinction, during which a significant proportion of the Earth's living species are killed off.

Palaeontologists have generally interpreted these mass extinctions ('missing links') as due to changes in climate or geography. But as long ago as 1950 Schindewolf put forward the suggestion that the spectacular extinctions at the end of the Palaeozoic era (about 230 million years ago) were caused by increased cosmic radiation onto the Earth's surface. A number of other extra-terrestrial hypotheses were suggested,

but they were not taken seriously due to lack of scientific evidence.

In 1980 Professor Alvarez and colleagues reported on the abnormal enrichment in iridium and other noble metals in terrestrial clay layers at the Cretaceous/Tertiary boundary in the Geological Column about 65 million years ago. This was thought to be due to the impact of meteorites on the Earth's surface, which caused an expulsion of terrestrial dust on such a scale that sunlight was blotted out for several weeks or months, causing the simultaneous extinction of plankton, of dinosaurs and many other groups. This hypothesis generated great excitement and still remains controversial.

Studies reported in 1984 by Raup and Sepkoski have detected a statistically highly-significant 26-million-year cycle in mass extinctions since the late Permian era (about 230 million years ago). This cyclicity hypothesis has been independently supported by David and colleagues (1984), who postulate the existence of an unseen companion-star to the Sun, occupying a highly eccentric orbit. At certain orbital positions, it is brought into the dense inner region of a comet cloud. By disturbing the cometary orbits it initiates an intense comet shower, leading to a series of terrestrial impacts lasting up to a million years.

Other models are based on long-term changes in cosmic radiation due to the Sun's oscillation about the galactic plane. Or extinctions might have been caused when the

Craters left by the impact of meteorites are scattered over the Earth's surface. On the left is a photograph of the Quebec Crater Lakes in Canada, taken by the Landsat space satellite. Was an increase in meteorite impacts responsible for the extinction of the dinosaurs?

On the right is Meteor Crater in Arizona. This crater was formed about 20,000 years ago. It is about 1,480 feet deep from the top of its raised rim to its floor, and about three-quarters of a mile across. One can imagine the amount of soil and dust it must have displaced.

Earth passed through interstellar gas or dust clouds, or intercepted a cometary shower.

CHANGES IN CLIMATE

Other explanations are more down-to-earth. Changes in sea-level and/or in temperature have been dramatic throughout time. Could these explain the mass extinction of plant and animal organisms on Planet Earth? These sharp peaks in the number of vanishing species have affected terrestrial as well as marine organisms, but the fossil record documenting their occurrence is more abundant in the marine realm. During geologically-brief intervals of several million years some of these events have eliminated most of the species and as many as half the families in the ocean. Devastation of this magnitude could have been inflicted only by significant changes in the environment on a regional or even a global scale.

Not all the explanations suggested for this are exotic. Some have looked to drastic changes in the environmental 'limiting factors' such as temperature and living space on the sea-floor. These ordinarily determine the distribution and abundance of species in the sea.

The most important factor limiting the geographical distribution of animal and plant species in the oceans seems to be water temperature. A particular species can survive only within a certain range of temperatures. Living animals confirm this. The geographical spread of a species is marked by temperature. Dr Steven Stanley provided new evidence in 1984 pointing to climatic cooling as the primary cause behind most of the known marine extinctions. He provides three types of evidence:

● In some cases **there is independent indication, such as the presence of glacial gravels, that cooling occurred at the same time as a mass extinction.** (Cooling need not always have been accompanied by glaciation, however.)

● The fossil record suggests that **species adapted to warm water or to a narrow temperature range have tended to suffer most,** as one would predict if the crises were brought about by a drop in temperature.

● The fossil record also indicates that **most marine mass extinctions were gradual rather than sudden events,** taking place over hundreds of thousands or even millions of years. This pattern is compatible with the usually slow and episodic rate of global cooling.

Probably the best-known mass extinction was the disappearance of the dinosaurs and other animal species at the

Numbers of species of marine life have been extinguished over relatively short periods at certain times of Earth's history. Were these extinctions caused by marked changes in sea temperature?

end of the Cretaceous Period, some 65 million years ago. This now seems to have been gradual. Stanley (1984) proposes that this crisis was not a single brief event. He suggests that different groups of organisms declined and became extinct at different times, over a period of at least two million years. The sequence is also significant, because it contradicts the ideas proposed by Professor Alvarez, that extinctions might begin at the bottom of the food web and propagate upwards in a kind of domino effect. Stanley's work shows that lower plants (mainly plankton) suffered at the very end of the Cretaceous crisis, after the decline of many plankton-eating organisms and after the total disappearance of the carnivores.

Professor Tony Hallam reported in 1984 that many problems and uncertainties remain and there is much to be learned about the response of particular groups of organisms and whole ecosystems to changes in physical environments. The extinction of plankton at the end of the Cretaceous Period, for instance, is still an enigma. Hallam remarks that before astronomers indulge in further speculations about the cause of mass extinctions of organisms on the Earth, they would do well to learn something about the rich strati-graphical record of their own planet.

Geologists have good evidence to correlate mass extinctions with a number of crises through the geological column at different times right back to the first event known to have taken place about 650 million years ago, late in PreCambrian times. But there does not appear to be a clear or agreed explanation of the reason for mass extinctions on the Earth. Although continued research into the fossil record should reveal further evidence of any firm link between cooling of the Earth's oceans and marine extinctions, Stanley suggests that it may never be possible fully to understand why these climatic changes happened millions or hundreds of millions of years ago.

Life from Space

I n 1903 Arrhenius introduced the 'Panspermia' theory, in which he visualized that life could have been transmitted to Earth by means of spores carried by meteorites.

Arrhenius wrote:

'On the road from the Earth the germs would for about a month be exposed to the powerful light of the Sun, and it has been demonstrated that the most highly refrangible rays of the Sun (ie ultra-violet) can kill bacteria and their spores in relatively short periods. As a rule, however, these experiments have been conducted so that . . . the spores were in a condition to germinate. That, however, does not at all conform to the conditions prevailing in planetary space. For Roux has shown that anthrax spores, which are readily killed by light when the air has access, remain alive when the air is excluded . . . All the botanists that I have been able to consult are of the opinion that we can by no means assert with certainty that spores would be killed by the light rays in wandering through infinite space.'

Various themes have been played on this extra-terrestrial origin-of-life motif in the intervening years. The early theories and ideas about molecules and life in space were not based on scientific observation but mainly on educated guesses and scientific intuition. In the last two decades, modern techniques of radio astronomy have shown that some of the possible precursors of 'life molecules' (such as methane, ammonia, formaldehyde and possibly simple amino-acids) are created in space by the action of ultra-violet light.

VIRUSES FROM SPACE?

During the late 1970s and early 1980s Sir Fred Hoyle — a famous British cosmologist and mathematician cum

Professor Svante Arrhenius, from Sweden, who in 1903 introduced the idea that life arrived on Earth from outer space.

science-fiction writer — has resurrected with his collaborator Chandra Wickramasinghe a modern version of the 'panspermia' theory. In their hypotheses, Hoyle and Wickramasinghe suggest that there is continuing recruitment of the terrestrial population of pathogenic bacteria and viruses from space. In particular they think that cometary tails may constitute an environment in which the microbes could be preserved and from which they could rather readily be transported to Earth. Hoyle thinks they are still arriving.

In a paper by Sir R. E. O. Williams to the Royal Society in 1981, the author's remit was to look at the patterns of a few infective diseases to see whether the gaps in knowledge are so profound as to justify involving such extra-terrestrial recruitment to our stock of pathogenic microbes. Or could the known and scientifically-tested facts suggest a far simpler explanation? In his conclusion Williams stated:

'From an Earth-bound view, the postulate does, however, seem unnecessary; there are good opportunities here for new strains, and surely new species, to evolve, and much of the 'evidence' called to deny the more conventional modes of spread ignores the fact that one must actually test for the microbes

CHARIOTS OF THE GODS?

Erich von Däniken's books have been read by millions. Not surprising, since they are very exciting stuff. Take this extract from the introduction to *Chariots of the Gods*:

'The gods of the dim past have left countless traces which we can read and decipher today for the first time because the problem of space travel, so topical today, was not a problem, but a reality to the men of thousands of years ago. For I claim that our forefathers received visits from the universe in the remote past. Even though I do not yet know who these extra-terrestrial intelligences were or from which planet they came, I

nevertheless proclaim that these "strangers" annihilated part of mankind existing at the time and produced a new, perhaps the first, Homo sapiens.'

In later books von Däniken upholds his view that spacemen visited the Earth long ago. He also suggests that many ancient objects were left by the 'gods of the dim past'. In *Gold from the Gods* he describes how, accompanied by an archaeologist, he allegedly descended into an astonishing world of subterranean galleries hidden beneath the tropical jungle of the Ecuadorian province of Moreno-Santiago. Metal tablets found in these

caves were claimed to support his views of previous visits from space. The book gives details of the underground furnishings, wall carvings, metal plaque library, birds, snails and crabs, conjuring up a remarkable scene.

These and other spectacular claims aroused world-wide interest, and the German magazine *Der Spiegel* sent an expedition to follow von Däniken's journey through South America and to interview the people mentioned in the book. The findings of this expedition repudiated many of von Däniken's assertions. According to his Ecuadorian guide, von Däniken did not actually descend into the themselves (or antibody to them) to determine the reservoirs and modes of spread of microbes.'

In later publications by Hoyle and Wickramasinghe, they argue that the evolutionary course of life on Earth has always been subject to cosmic influence and that the biological make-up of living organisms on Planet Earth is radically changed by the arrival of new genes from outer space. They claim in their hypothesis that these ideas provide an explanation for the sudden bursts of new life-forms and other dramatic developments that occur. This hypothesis has been further extended and made more bizarre in that Hoyle and Wickramasinghe now attribute not only the origin of terrestrial life but its continuing evolution to cosmic influences.

This has all been received very sceptically by the scientific world and is not in any way seriously accepted. Because of the nature of the topic, these hypotheses are based on very limited scientific data or results; they are developed from much inference and extrapolation. The original cosmochemical observations provided much interesting information on the occurrence and nature of organic molecules and macromolecules in outer space, but their recent foray into explaining the origin of life on Earth and

galleries at all, but rather questioned the guide for all the information he had. He later admitted that his itinerary in Ecuador did not allow him the time necessary even to reach the location so vividly described in his book.

Many professional scientists and religious people have examined and criticized the material in von Däniken's books. The unanimous conclusion is that his claims are totally false; they have no scientific or archaeological support. One of the kindest comments on his books is that they 'are not written to persuade the informed reader. They are a romanticist's fiction, and to examine or criticize the

material at any great length seems pointless'.

And yet these notions of 'gods from space' playing a decisive part in prehistory and human evolution have enjoyed a tremendous vogue. They obviously touch off some deep response in a great many people. Von Däniken himself believes that one of the major reasons for the success of his books (over 12 million sales) is religious uncertainty. For those many people in the world today who need certainty, assurance and stability in a rapidly-changing society, von Däniken's books have the attraction of being readily understood, vividly presented and professionally written, with

the impression of strong conviction. The books provide a substitute for faith and reason in the shape of fantasy and escapism. Unfortunately the escape is only brief, because the books are almost pure fiction and vulnerable to critical demolition.

Yet many religious cults are alive and practised today whose origins were equally shaky. People are searching to find their true Creator and a meaning for life. The Bible directs all such honest enquirers to God the Creator and to Jesus his Son.

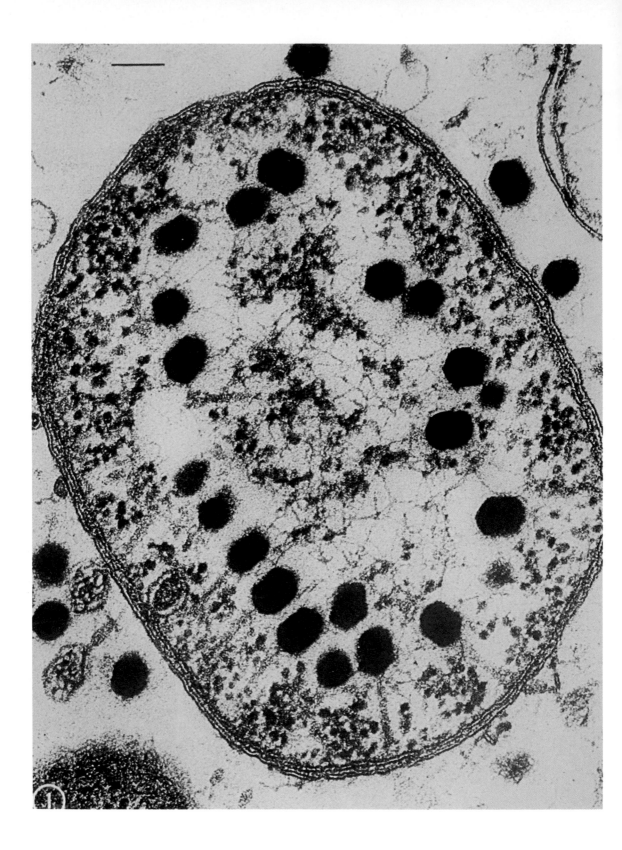

This electron microscope picture shows T2 bacteriophage viruses attacking and entering into a cell (of an Escherichia Coli **bactumium). Hoyle and Wickramasinghe believe that viruses reached the Earth from space.**

evolution by cosmic influences is not supported by the vast majority of scientists.

As we saw earlier, the evidence is against spores and remnants of living organisms as constituents of meteorites. But analyses do show that some chemical precursors of life exist in outer space and can be transported through space by carbonaceous chondrites. It is interesting to speculate that certain important chemicals (probably precursors of molecules required for life) that exist in space and in carbonaceous chondrites *could have been* distributed by such objects on landing on Earth some 4,000 million years ago and *may have been* the precursors of primitive life. Careful studies on interstellar molecules, organic components in meteorites and very old PreCambrian rocks may yet help us with the mysteries of the origin and development of life on Planet Earth.

Nobel prize-winner Francis Crick and Professor Leslie Orgel have considered the possibility that life is deliberately transmitted to chosen planets by advanced civilizations elsewhere in the Universe. They suggest that there are no insuperable technological obstacles to this, and add cynically that Planet Earth should soon be able to pollute other planets in the same way!

Molecules in Interstellar Space

T he vast space between the stars is filled with matter. Most of it is hydrogen and helium, which under interstellar conditions cannot form solid bodies — even at minus 270°C (3° Kelvin) the Universe is simply too hot for them to condense. Yet there are solids in interstellar space: the tiny frozen particles called interstellar grains.

These grains form from heavier chemical elements synthesized by thermonuclear fusion in stars and supernovas. The most abundant of these elements are oxygen, carbon and nitrogen (known collectively as the 'organics'). Magnesium, silicon and iron are next in abundance. The interstellar grains consist of these six condensable elements together with hydrogen which is captured from the surrounds.

Astronomical and laboratory-simulation studies have shown that interstellar grains have a remarkable complexity; they are not inert, amorphous lumps of cosmic dust as previously thought. It appears that a typical interstellar grain has a distinctive internal structure, with a core made of silicates (inorganic sand-like material) and an outer coating of more volatile organics. This outer surface of the grain, on which the organics accrete, is where complex chemical processes appear to take place.

These interstellar grains are not the only molecular structures in space. The spectra from hot-emission nebulae have been analyzed in some detail by optical spectroscopy. But it is the radioastronomers who have succeeded since 1968 in identifying over fifty molecules in dense concentrations of the interstellar gas now generally termed

Radiotelescopes are essential tools for probing outer space. They locate distant radio sources, and also search for interstellar molecules. This Australian CS1RO 210-foot radiotelescope is at Parkes, New South Wales.

These plots of the spectra of the compounds Formamide (A) and Methanimine (B) are from very different sources. A1 and B1 are laboratory recordings of their spectra, but A2 and B2 are radiotelescope signals from Sgr B2 stars, showing that formamide and methanimine are present there. Note the close correspondence in the two pairs of plots. (Data provided by Professor R. D. Brown of Monash University, Australia.)

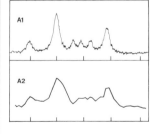

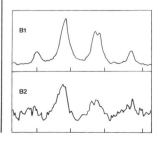

molecular clouds. From analysis of rotational radio and microwave emission from molecules it has been shown that the vast black clouds which litter the space between the stars are full of compounds.

Although it is becoming increasingly evident from radio-astronomy studies that interstellar space contains many and varied molecules, some of remarkable complexity, it is not at all easy to identify the molecules. In the laboratory a spectrum is obtained by passing monochromatic microwave radiation through a cell containing the vapour of the molecule to be studied. From these laboratory experiments, very precise frequencies characteristic of the molecule are used to obtain its spectrum. No two molecules show the same spectroscopic pattern, so these laboratory-derived molecular spectra can be used to 'scan' the galaxies to monitor their occurrence.

Radio telescopes are essentially glorified radios with large, steerable, highly-directional aerials. Interstellar space can be searched for molecules by tuning the radio telescope to the same frequency as that determined by the laboratory study, pointing the telescope at a suitable interstellar source and analyzing the weak incoming signal to test if a molecule is emitting the same frequency (or preferably set of frequencies).

IDENTIFYING MOLECULES IN SPACE

Owing to the high cosmic abundance of hydrogen, H_2 is by orders of magnitude the most common molecule in the interstellar gas; in both number-density and mass the other molecules rank merely as trace constituents.

The known interstellar molecules consist of simple inorganic compounds, stable organic molecules and unstable compounds: radicals, ions, isomers and polyunsaturated carbon chains. Interstellar molecules are as a rule composed of only five elements: hydrogen, carbon, nitrogen, oxygen and sulphur. The highest molecular-weight and most complex interstellar molecules are unstable polyene-type with relative high molecular weights. None of the inorganic molecules in space contains more than four atoms, so as on Earth it is apparently the carbon-bond that is the key to the synthesis of complicated molecules.

Astronomers have carried out observations right inside the galaxy Sagittarius A, where the gas density varies

SOME MOLECULES IN INTERSTELLAR SPACE

INTERSTELLAR INORGANIC MOLECULES

diatomic	triatomic	tetra-atomic
H_2	H_2O	NH_3
CO	H_2S	
CS	SO_2	
NO	HNO	
NS	OCS	
SiO		
SiS		

INTERSTELLAR ORGANIC MOLECULES

alcohols
CH_3OH	methanol
CH_3CH_2OH	ethanol

aldehydes and ketones
H_2CO	formaldehyde
CH_3CHO	acetaldehyde
H_2CCO	ketene

acids
HCN	hydrocyanic
HCOOH	formic
HNCO	isocyanic

amides
NH_2CHO	formamide
NH_2CN	cyanamide
NH_2CH_3	methylamine

esters and ethers
CH_3OCHO	methyl formate
$(CH_3)_2O$	dimethyl ether

sulphur
H_2CS	thioformaldehyde
HNCS	isothiocyanic acid
CH_3SH	methyl mercaptan

paraffin derivatives
CH_2CN	methyl cyanide
CH_3CH_2CN	ethyl cyanide

acetylene derivatives
HCCCN	cyanoacetylene
$HCCCH_3$	methylacetylene

others
CH_2NH	methylenimine
CH_2CHCN	vinyl cyanide

INTERSTELLAR UNSTABLE MOLECULES

radicals	ions	isomers	carbon chains
CH	CH +	HNC	HC_5N
CN	HOCO + ← or → HOCN		HC_7N
OH	HN $_2^+$		HCN_9N
SO	HCS +		
C_2	HCO +		
HCO			
C_2H			
C_3N			
C_4H			

This list of some of the groups of molecules and their formulae, is taken from P. Thaddeus, 1982.

appreciably but rises towards the galactic centre. About 750 light years from the centre there is a marked peak. Here the gas is concentrated into a clumpy ring of giant gas and dust clouds, relatively cool and rich in interstellar molecules of all kinds. As well as fairly common varieties such as hydroxyl (OH) and carbon monoxide (CO), the clouds contain many more complex molecules. Astronomers have detected formaldehyde (HCHO), formamide ($HCONH_2$), acealdehyde (CH_3CHO) and ethanol (CH_3CH_2OH).

INTERSTELLAR GRAINS

Structure

Each grain begins as a silicate core and around it ice forms. Into the grain are absorbed organic molecules or radicals which can react and recombine to give more complex molecules. Ultra-violet radiation is thought to be involved in these reactions.

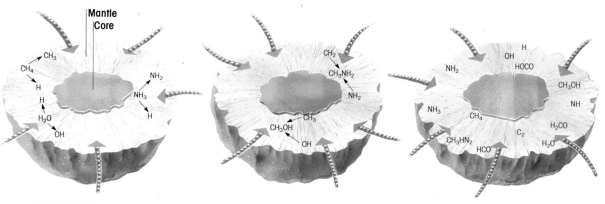

Different types

Grains appear to have different layers depending on their stage of development and the different clouds or incipient stars in which they are formed. Grains formed in a diffuse cloud (**A**) have a silicate core and 'organic' layer. Grains in a molecular cloud (**B**) have in addition an outer mantle consisting of ice condensate produced by ultra-violet radiation. Grains in a dense cloud which is about to form a star (**C**) have a more complex structure.

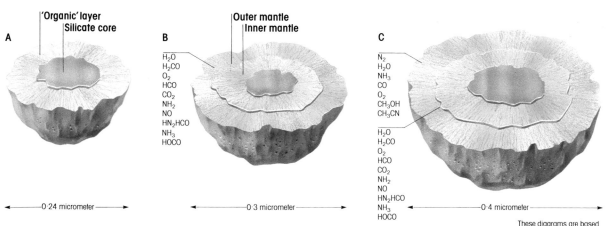
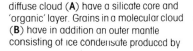

A ── 0.24 micrometer ──

B ── 0.3 micrometer ──

C ── 0.4 micrometer ──

B	
H_2O	
H_2CO	
O_2	
HCO	
CO_2	
NH_2	
NO	
HN_2HCO	
NH_3	
HOCO	

C	
N_2	
H_2O	
NH_3	
CO	
O_2	
CH_3OH	
CH_3CN	
H_2O	
H_2CO	
O_2	
HCO	
CO_2	
NH_2	
NO	
HN_2HCO	
NH_3	
HOCO	

These diagrams are based on work by J. M. Greenberg

Within these identified organic molecules there appears to be a particular structural theme. All interstellar molecules appear to have simple, linear, heavy-atom backbones; branched chain and rings have not yet been observed in interstellar molecules. Carbon-sulphur and carbon-ring structures have been identified, however, in carbonaceous chondrites, and these organic ring compounds are considered to have formed by condensation on the mineral grains of the chondrite.

The origin and significance of interstellar molecules has been the topic of much recent research. This has provided various different interpretations. In a review of interstellar molecules published in 1984, Dr H. Kato wrote:

'Suffice it to say that interstellar studies have shown that some very long molecules exist in the space between the stars. They may be very long indeed and their relationship with interstellar dust is far from being understood. In fact, it is only now that a possible relationship could even be contemplated. The long chains may be an intermediate form of carbon between the well-known small species consisting of one, two or three carbon atoms and particles with a high carbon content such as soot.'

Another possibility is that these molecular clouds are the raw material out of which stars and particularly planets form. The mechanism of planet formation is still far from clear, and the relationship of the molecules in these clouds with those in Earth's living environment is even less clear. We now know that molecules are formed in stars and pushed out into space.

There are therefore now three ways that have been put forward in which bioemotive molecules, such as glycine, can be formed:

● **in Earth's living environment** by Urey-Miller type processes (where forces such as lightning provide the energy for the formation of small biomolecules);

● **in the cold interstellar clouds** by ion-molecule reactions and perhaps also by grain catalysis;

● **in circumstellar shells.**

There appears to be an inexorable drive to form the molecules that are the building-blocks of life.

The new results herald a fresh look not only at the origin of the living environment but also at the mechanism of grain formation and at grain identity, as well as the formation of larger objects such as planets.

Science and Creation

T he topic of the origin of life has in recent years become of wide interest to the whole of humanity. The subject is not only of interest to scientists, but also to theologians, philosophers, indeed to every human being. It attracts the attention of scientists in different disciplines and brings them together from different countries.

The history of the Earth covers an enormous time-span, and the data on which it is based have very often been rather scanty. As a result, geologists have tended to concentrate most of their studies on those periods of the Earth's history which are relatively well-known, and have sometimes neglected the more obscure periods involving the oldest rocks. This means that geology has supplied us with a fair amount of data about fossilized remains of earlier life-forms extending over about the last 600 million years — from the Cambrian Period to the present day. But even in this more 'recent' period, the palaeontological record is still very incomplete in most areas, and so concentrated studies have had to be devoted to the development of life during its more recent history. Geological data about the origin of life, which needs to reach far back into the PreCambrian about 3,000–4,000 million years ago, is much more fragmentary.

Occasional studies during the early part of this century identified evidence for PreCambrian life, but this was not accepted by many scientific sceptics of the day. It was not until 1954, when Elso Barghoorn and Stanley Tyler investigated the microfossil structures in the 2,000-million-year-old

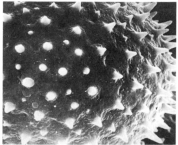

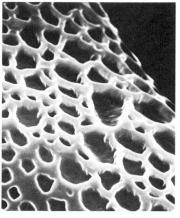

This beautiful present-day flower will disappear, but its pollen and spores may remain. The most abundant plant fossils found in ancient rocks are pollen and spores. These tiny objects are preserved because of their highly-resistant outer coat of sporopollenin. Modern pollen grains Malva sylvestis (top), Billbergia nutans (middle) and Alnus glutinosa (bottom) all have inert sporopollenin walls; some of these grains will be preserved in sediments. Every plant pollen or spore is a different size, shape and surface design, and can be specifically identified in nature and in sediments.

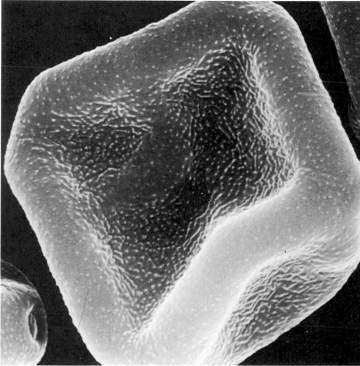

Gunflint Cherts near Lake Superior, that it became generally accepted that there is evidence for life in PreCambrian rocks, of different age from the four corners of the Earth. They have shown that fossil remains occur throughout the PreCambrian back to about 3,400 and possibly 3,700–3,900 million years ago. We have attempted to evaluate this evidence for life in the PreCambrian, but it still remains an important fact that there are incomparably more geological data available about the development of life in the last 600 million years than about its earlier period. And about life's origin and the possible period of transition from non-living to living some 3,400–3,900 million years ago the rocks give us only the faintest clues.

GENESIS AND ORIGINS

One great belief unites all Christians as they think about the beginnings of life: that God is the Creator of our Universe and our planet. From this belief we draw the conclusion that the Universe is not random but purposeful. It fills us with awe and gratitude to our Creator.

As we read the first chapter of Genesis in the Bible we find that it gives a true account of God's creation of the Universe, of the Earth, of living things and of mankind in his place among them. We respond gladly to the writer's assertion about this creation that 'it was good'. The very many scientists who are Christians find nothing discordant between the Bible's account and a scientific account, provided, as we shall discuss later, that both are kept in their proper places.

As to the way Genesis 1 should be interpreted, however, it is common knowledge that there are different views among Christians. At one end of the scale are some who believe that this chapter is to be understood completely literally, meaning that God created everything as it now is in six days some 4,000 years ago. Others hold that this is a wrong interpretation, and that the six 'days' are not days in which God created but days in which God revealed to mankind the truth about his creation; such an interpretation allows the time-scale for life's development to be as most scientists believe it to be. Others again believe that the Genesis writer's account of creation does not pretend to be scientific truth but truth in a poetic mode, giving us in a matchless story the essential theological realities that God is Creator and stands in relation to his creation as its sovereign Lord.

None of these positions is inherently impossible. God

This space photograph of the densely-populated coastline around New York and Long Island may make us ponder the phenomenon of human life on our planet. The development of higher life-forms and of our own species is not the theme of this book. But the existence of life itself is remarkable enough. Its arrival remains the subject of intense scientific investigation and discussion.

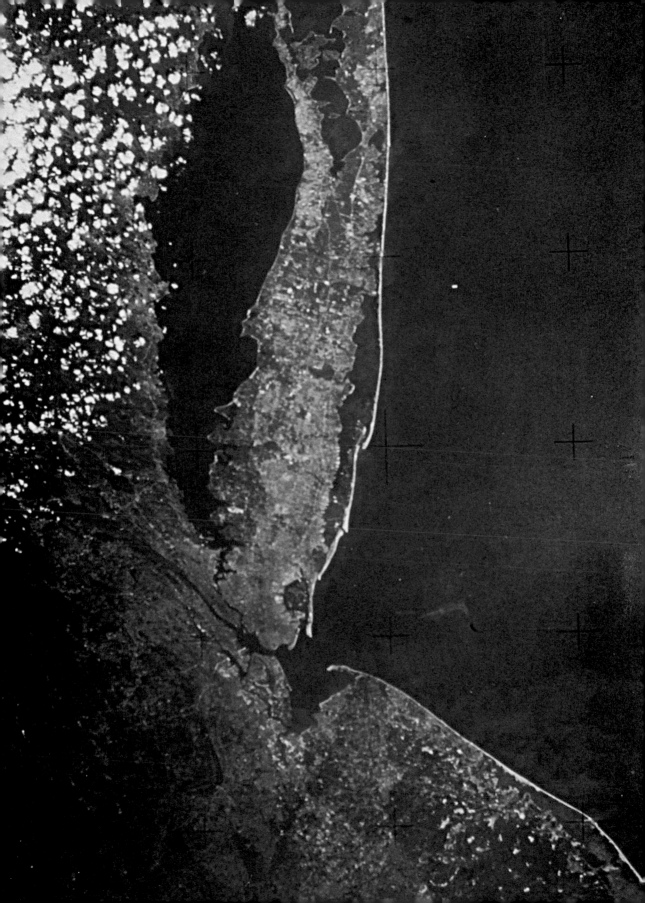

could have created a Universe relatively recently which gives the appearance of being very old. But those many Christians who do not take the 'days' of Genesis 1 as twenty-four-hour days find no difficulty in treating the scientific evidence as it stands. And when scientists who are Christians do this, they find much to suggest it is unlikely that life began by chance.

Many words have been spoken and written about science and the Christian faith, but the best witness is shown through the living faith of many, many scientists. A senior member of the US NASA Space Programme once visited a church in the NASA community, and said this: 'How comforting it is to see the highly accomplished men in the engineering and scientific community come to church and to Christ for answers for their lives that they alone do not have nor can explain.'

WHAT KNOWLEDGE DO WE NEED?

There is intense excitement to be found in following the scientific evidence back towards life's origins. But unfortunately some Christians are reluctant to join the search. Even among those who are prepared to get acquainted with the fossil history and the development of living systems, there is sometimes a real hesitancy to go back to the question of how and when life began.

THE GUNFLINT FORMATION

Probably the most important fossil-biology breakthrough in the 1950s was made by Tyler and Barghoorn in 1954. They discovered a rich and diverse mixture of microscopic sea-floor and plankton life-forms in stromatolite cherts of the approximately 2,000-million-year-old Gunflint Formation.

A remarkable outcropping of PreCambrian rocks along the shore of Lake Superior in Western Ontario shows a sequence of sediments known as the Gunflint Iron Formation.

Most of the Gunflint microfossils are three-dimensional, and many show detailed surface patterns. Elso Barghoorn proposed that the structure of these fossils has been preserved without distortion by being deposited in a siliceous solution that later crystallized into chert, much as modern biological specimens are preserved by being embedded in plastic.

The most abundant of the Gunflint microfossils are

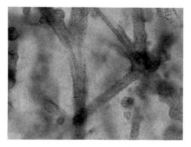

filamentous structures. These primitive plants have been assigned on the basis of their structure to four genera including five species. They resemble some present-day blue-green algae, particularly the filamentous Oscillatoria and related forms. Another fossil form resembles the modern iron-oxidizing bacterium Crenothrix.

There is much compelling palaeobotanical, geochemical and geological evidence to support the belief that microbial life was extant approximately 2,000 million years ago (the probable age of the Gunflint cherts). The Gunflint fossils are

Two reasons seem to lie behind this. One is a suspicion that to probe the 'natural' way life began is somehow to deny that God created it. I disagree with this. To say that God used means to bring life into existence in no way takes away from his creative activity, nor does it make life's origin any less miraculous; we rightly speak of 'the miracle of life'.

But the other reason why some Christians feel uneasy on this ground is that some of the early sallies into examining the origins of life were open to accusations of bias in an atheistic direction. This is something which lay people (and some scientists) do not often recognize or accept. The reason why the Soviet Union, under the initial lead of Academician Professor Aleksandr Oparin, were pioneers in studies on the origin of life was obviously not for purely scientific purposes. Such researches were carried out, and presumably still are today, expressly to further the influence of Marxist-Leninist doctrine. This purpose becomes even more apparent when one realizes that Professor Oparin's main British scientific colleagues and supporters, Professor J. B. H. Haldane (biophysicist) and Professor J. D. Bernal (crystallographer), were both, at least during part of their scientific careers, opposed to religious beliefs.

From the 1920s to the 1940s these studies deliberately

demonstrably indigenous to the stromatolite cherts in which they occur, and were deposited with them. The fact that this material was once living has been firmly documented as have its biological affinities. It is comparable in structure to specific modern microbes.

These fossils of micro-organisms were found in 2,000 million-year-old rocks of the Gunflint Formation on the shores of Lake Superior in Canada. They resemble some present-day blue-green algae. It was the fossil evidence in these Gunflint cherts that finally caused it to be widely accepted that there is evidence for life in PreCambrian rocks.

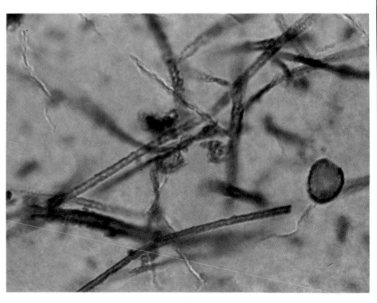

aimed to focus scientific thinking away from religion. Their aim was a completely materialistic, humanistic theory of life — not only of evolution but also of origins. The political and doctrinal purpose of these studies was not a pure scientific search for the truth about the origin and development of life, but specifically directed to doing away with creation altogether.

In recent years, the scientific enquiry into the origin of life has been much more objective and of high academic quality. The published work on scientific experiments in bio-chemistry, biology, geology and astronomy appears to be based on facts and mostly free from both political and religious bias. Some of the more recent scientific reports and interpretations may appear highly speculative, but they are all based on scientific experiment and scientific lines of reasoning.

Unfortunately, too few people are acquainted with science and its methods, and any confusion about creation and evolution appears to coincide with the apparent decline in the standard of science education in our schools. There are problems, especially for young people, in that our modern space-age world can appear to bring mankind closer to omniscience. Yet any first-class scientist will say that the more knowledge is acquired, the greater our awareness of the unknown.

This is the essence of wisdom. Mankind's knowledge is always relative to his ignorance not to infinite knowledge. Infinite knowledge belongs to God the Creator. None of our knowledge, known and to-be-discovered, can compare with the knowledge of God. But this should not lead us to believe that landing on the Moon, exploring the Solar System or enquiring about the origin of life is an intrusion upon the knowledge that belongs only to God the Creator. Our scientific and technological know-how and our spirit of enquiry are God-given: mankind's quest to understand his origin and development is a response to the life breathed into us by our Creator.

There is no conflict between scientific knowledge about the Earth, Solar System or Universe and personal knowledge of God. Professor Boyd, in his book *Can God be Known?*, points out that when we speak of knowledge it is in at least three senses: mathematical, scientific and personal. He shows how each type of knowledge is similar in that it proceeds from presuppositions which are eminently sensible but cannot be demonstrated. Scientific knowledge requires

This celebrated photograph of 'Earth-rise' as seen from the Moon reminds us that the search for the origins of the Universe and of life now reach beyond our own planet. It also reminds us, as it reminded some of the American astronauts, that questions about origins have never been purely scientific questions. They are religious ones as well. Did it all happen by chance, or is there purpose in the Universe?

that we assume the existence of the external world and the uniformity of nature. Personal knowledge requires that we assume other minds and personalities like our own.

Scientific knowledge gives us an I-It relationship with the natural world, while personal knowledge depends on an I-You encounter with other people. Science is concerned with description; religion with encounter. Science can be viewed as learning about God by studying his creation, and many scientists who are Christians see their work in this way; Christianity is a personal relationship with and personal response to the Mind behind the Universe and Life.

In accepting the truth of a Universe where Planet Earth and the other planets move around the Sun, in appreciating some of the vast activities and events that are going on at the edge of a galaxy where there are millions of other suns with their planets and moons, the believer is moved to praise God for the greatness of this creation. The knowledge of the Big Bang and the expanding Universe, of the origin of life and its meaning may just set our minds buzzing, or it may suddenly wake us up to appreciate 'The Mighty One, God the Lord!'

A SELECTION OF HELPFUL BOOKS AND PAPERS

Acworth, R. (1970) *Creation, Evolution and the Christian Faith* London; Evangelical Press

Barghoorn, E. S. (1971) *The Oldest Known Fossils* Scientific American, 224, 30–42

Bracewell, R. N. (1975) *Intelligent Life in Outer Space* San Francisco; W. H. Freeman

Brooks, J. (1981) *Organic Matter in Meteorites and PreCambrian Rocks: clues about the origin and development of living systems* Phil. Trans. R. Soc., London, A303, 595–609

Brooks, J. and Shaw G. (1973) *Origin and Development of Living Systems* London and New York; Academic Press

Brooks, J. and Shaw G. (1978) *A Critical Assessment of the Origin of Life*, in Noda H. (ed) *Origin of Life* Tokyo; Japan Scientific Societies Press

Calder, N. (1972) *Restless Earth* London; BBC Publications

Calvin, M. (1969) *Chemical Evolution* Oxford; Clarendon Press

Clark, R. E. D. (1958) *Creation* London; Tyndale Press

Cloud, P. (1978) *Cosmos, Earth and Man* New Haven and London; Yale University Press

Day, W. (1979) *Genesis on Planet Earth* East Lansing, Michigan; The House of Talos Publishers

Disney, M. (1984) *The Hidden Universe* London; J. M. Dent

Filby, F. A. (1977) *Creation Revealed* London and Glasgow; Pickering and Inglis

Folsome, C. E. (1979) *The Origin of Life* San Francisco; W. H. Freeman

Gass, I. G., Smith, P. J. and Wilson, R. C. L. (1971) *Understanding the Earth* Milton Keynes; The Open University Press

Gribbin, J. (1981) *Genesis: The Origins of Man and the Universe* London; J. M. Dent

Green, M. (1968) *Runaway World* London; Inter-Varsity Press

Hallam, A. (1983) *Great Geological Controversies* London; Oxford University Press

Hoyle, F. (1983) *The Intelligent Universe* London; Michael Joseph

Hoyle, F. and Wickramasinghe, N. C. (1978) *Lifecloud: The Origin of Life in the Universe* London; J. M. Dent

Hoyle, F. and Wickramasinghe, N. C. (1981) *Evolution from Space* London; J. M. Dent

Hutchinson, R. (1983) *The Search for our Beginning* London; British Museum (Natural History) and Oxford University Press

Institute of Geological Sciences Series, published by Her Majesty's Stationery Office, London:
The Age of the Earth (1980)
Moon, Mars and Meteorites (1977)
Britain Before Man (1978)
Volcanoes (1977)
The Story of the Earth (1977)

Lambert, D. (1979) *The Earth and Space* London; Grinswood and Dempsey

Lovell, B. (1979) *In the Centre of Immensities* London; Hutchinson

Lovelock, J. E. (1982) *GAIA — A New Look at Life on Earth* London; Oxford University Press

Noda, H. (1978) *Origin of Life* Tokyo; Japan Scientific Press

Oparin, A. I. (1968) *Genesis and Evolutionary Development of Life* London and New York; Academic Press

Parrott, B. W. (1969) *Earth, Moon and Beyond* Waco, Texas; Word Books

Peacocke, A. R. (1971) *Science and the Christian Experiment* London; Oxford University Press

Ponnamperuma, C. (ed) (1977) *Chemical Evolution and the Early PreCambrian* New York and London; Academic Press

Royal Society Discussion Meeting (1982) *Molecules in Interstellar Space* London; Royal Society

Rutten, M. G. (1971) *The Origin of Life by Natural Causes* Amsterdam and London; Elsevier

Ryan, P. and Pesek, L. (1978) *Solar Systems* London; Allen Lane

Schaeffer, F. A. (1973) *Genesis in Space and Time* London; Hodder and Stoughton

Schopf, J. W. (ed) (1983) *The Early Biosphere* Princeton University Press

Spanner, D. C. (1966) *Creation and Evolution*
    London; Falcon
Taylor, G. R. (1982) *The Great Evolution Mystery*
    London; Secker and Warburg
von Ditfuth, H. (1981) *The Origins of Life* San
    Francisco; Harper and Row
Weinberg, S. (1983) *The First Three Minutes*
    London; Flamingo/Fontana
Wiseman, D. J. (1977) *Clues to Creation in Genesis*
    London; Marshall, Morgan and Scott

Interested readers should follow scientific
research, developments and reports in the
various serious scientific journals. They are
especially recommended to read *Nature*,
Scientific American and *New Scientist*, which
publish many relevant and interesting articles
and reviews on subjects related to the origin
of life.

ACKNOWLEDGMENTS

I am deeply grateful to many, scientists and non-scientists, who have joined in discussion on the origins of life. Long may such interesting and informative discussion continue. However, the views expressed in this book are solely my own. Also, the interpretations of other people's work are sometimes my own, and it should not be inferred that the quoted authors necessarily agree with my interpretations.

I want to thank Aileen Henderson, Margo Robertson and Jane Russell for typing (and retyping) the manuscript. And my wife Jan and children Daniel and Naomi have given constant support and encouragement.

The photographers and organizations below have kindly agreed to the use of their photographs. (I am particularly grateful to colleagues in the relevant scientific fields who loaned me their own photographs for publication.)

Prof W. G. Chaloner 16, 22 (both), 24, 25 (above left and below), 32
Prof J. Collinson 20 (above), 26 (below), 28, 89 (above), 98, 109, 117
CSIRO Australia, Division of Radio Physics 141
Daily Telegraph Colour Library 131
Earth Resource Foundation NSW 26 (above), 27
Dr G. Farrow 133
M. J. Fisher 25 (above right)
Prof D. O. Hall 113 (above)
Hulton Picture Library 79
Institute of Geological Sciences, reproduced by permission of the director, British Geological Survey, crown copyright reserved 76–77, 119 (all), 123
Marathon Oil UK Ltd 26 (below)
Dr S. Moorbath 114

Dr Marjorie Muir/J. Brooks 107 (all), 113 (below), 127 (all), /P. Grant 106 (all), 108 (top two), 111, 112, 113 (middle), 150, 151
NASA 6, 10, 41, 46, 50–51, 62, 63, 68, 69, 70, 71, 72, 73, 122, 149, 152
Natural History Museum, London 125
Dr H. Pflug 116
Jean-Luc Ray 146
Royal Astronomical Society 124
Science Photo Library 3, 89 (below), 92, 95, 100–101, 130, 138, 147 (all)
Svenskt Pressfoto 135
Prof Janet Watson 105
ZEFA 18–19, 30, 31 (both), 33 (above), 36–37, 43, 67, 90–91, end papers

Fred Apps composed the illustrations on pages 12–13, 17, 22–23, 33, 39, 47, 54–55, 56–57, 60–61, 80, 143

GLOSSARY

Abiological/abiotic Without life. Often used to describe situations where life has played no part in the end result or product.

Aerobic and Anaerobic With and without air respectively. Used by scientists to describe environments which are respectively surfeited with or deficient in oxygen.

Amino-acid A compound containing an amino (NH_2) and a carboxyl (COOH) group. There are twenty-two amino-acids, which in specific and various combinations make up all known proteins.

Andromeda Nebula The large galaxy nearest to our own. A spiral containing about 3×10^{11} solar masses.

Asteroids Small bodies in orbit around the Sun. Their main planet-like orbit is between Mars and Jupiter.

Atom The smallest unit of an element, consisting of protons, neutrons and electrons.

Big-Bang cosmology The theory that the expansion of the Universe began at a finite time in the past, in a state of enormous density and pressure.

Carbon-14 Radioactive isotope of the element carbon, which has a half-life of 5,730 years. Used to date events back to about 50,000 years ago.

Chlorophyll Green pigment which makes it possible for plants to synthesize food.

Compound A combination of atoms (or ions) of different elements. The way in which they are combined is called a bond.

Continental Drift The theory that earlier continents split into pieces that drifted laterally to form the present-day continents.

Continental Shelf A shallow, gradually-sloping zone extending from the sea margin to a depth at which there is a marked or steep descent into the depths of the oceans down a continental slope.

Cosmic dust Very fine material that falls on Earth from space.

Cosmic rays The nuclei of atoms that reach the Earth from space with enormous energy because of their great velocity.

Cosmochemistry Chemistry of the Universe.

Cosmology Science of the Universe.

Crust The outermost layer of the Earth, defined by seismic waves and by its chemistry. Composed of solid rock, some 20–40 miles thick on the continents and about 3–5 miles thick under ocean basins.

Deformation (of rocks) Any change in the original shape or volume of a rock mass. Folding, faulting and plastic flow are common modes of rock deformation.

Deuterium A heavy isotope of hydrogen (H2).

DNA Deoxyribonucleic acid, a long helical macromolecule that contains coded genetic information, which determines the hereditary characteristics of living organisms.

Doppler Effect A change in the observed frequency of a wave resulting from relative motion between observer and source.

Electron The lightest massive elementary particle. All chemical properties of atoms and molecules are determined by the electrical interactions of the negatively-charged electrons with each other and with atomic nuclei.

Element A unique combination of protons, neutrons and electrons that cannot be broken down by ordinary chemical methods.

Erosion The removal (or transportation) of material by abrasion, weathering and similar processes.

Eukaryotic cells Biological cells which have nuclear walls and the capacity for reproduction by replication of DNA.

Fossil Geological evidence of past life, such as the impression of a leaf in a rock, a pollen or spore, the footprint of a long-extinct animal or the shell of an ancient organism.

Fossil fuels (petroleum, coal) Fuels consisting of organic remains.

Frequency The rate at which crests of any sort of wave pass a given point.

Galaxy A large gravitationally-bound cluster of stars, containing up to 10^{12} solar masses. Our galaxy is often called The Galaxy — but other galaxies are usually classified according to their shape (spiral, for example, or elliptical).

Gene The basic hereditary unit of living organisms.

Geochemistry The chemistry of the Earth.

Geological Column A chronological arrangment of rock units in sequence with the oldest units at the bottom and youngest at the top.

Geology An organized body of knowledge, information and experiment about the Earth. Often equated with Earth Sciences. (Historical geology deals with the history of the Earth, including a record of life on Earth as well as physical changes in the Earth itself.)

Geophysics The physics of the Earth.

Gondwanaland A Mesozoic land mass (continent) that split into present-day South America, Africa, Australia, India and Antarctica.

Gravitation The attraction of all matter in the Universe to all other matter.

Half-life The time needed for one half of a sample of a radioactive element to decay.

Helium The second lightest, and second most abundant, chemical element.

Hubble's Law The relation between the velocity at which moderately-distant galaxies are receding and their distance. Hubble's constant is the ratio of velocity to distance in this equation.

Hydrocarbon A compound of hydrogen and carbon that burns in air to form water and carbon dioxide. There is a large series of hydrocarbons, the first member of which is methane (CH_4). Petroleum is mainly a complex mixture of hydrocarbons.

Hydrogen The lightest and most abundant chemical element.

Igneous rock An aggregate of interlocking minerals formed by the cooling and solidifying of magma.

Infra-red radiation Electromagnetic waves with a wavelength that is intermediate between visible light and microwave radiation.

Ion An electrically unbalanced form of an atom, produced by the gain or loss of an electron.

Isotopes Different forms of an element produced by variations in the number of neutrons in the nucleus and consequently in the atomic weight of the element. Isotope dating is determining the age of a body by measuring the ratio or ratios of the isotopes that it contains.

Laurasia The ill-defined ancient land mass (continent) from which the present North American and Eurasian continents have been derived.

Light year The distance that a light ray travels in one year. Equal to approximately 11,830 million million miles.

Lithosphere The outermost, rigid layer of the Earth, commonly about sixty miles thick. It is broken into moving tectonic plates and rests on the asthenosphere.

Magma A naturally-occurring silicate melt.

Mantle The intermediate zone of the Earth. Surrounded by the crust, it rests on the core at a depth of about 1,800 miles.

Metamorphic rock Any rock that has been changed in texture or composition by heat, pressure or both.

Meteor The streak of light emanating from a transient celestial body that enters the Earth's atmosphere with great speed.

Meteorite A stony or metallic extra-terrestrial body that has fallen to Earth from outer space.

Meteoroid Collective name for relatively small stony or metallic particles in outer space or arriving from outer space.

Milky Way The ancient name of the band of stars which mark the plane of our own galaxy. Sometimes used as a name for our galaxy.

Moon A natural satellite, as of a planet.

Nebula Extended astronomical object with a cloudlike appearance. Some nebulae are galaxies; others are actual clouds of dust and gas within our galaxy.

Neutron The unchanged particle present with protons in ordinary atomic nuclei.

Nucleus The protons and neutrons forming the central part of an atom. — The major organelle in eukaryotic cells that contains the replication DNA.

Ozone A highly poisonous and explosive blue gas. It is a rare form of oxygen in which three atoms instead of two are combined together.

Petrology The study of rocks.

Photosynthesis Process by which chlorophyll-containing plants synthesize food, in the form of carbohydrate, from carbon dioxide, water and light energy.

Plate tectonics The study of the motions and interactions of tectonic plates.

Primeval soup The accumulation which it is suggested took place of abiotically-synthesized chemicals in primitive oceans (or ponds) to give organic-rich soup or broth.

Prokaryotic cell An organism lacking a well-defined nucleus. (Examples: bacteria and blue-green algae.)

Protein Complex molecule made up of amino-acids.

Proton The positively-charged particle present with neutrons in ordinary atomic nuclei.

Radioactivity The spontaneous breakdown of an atomic nucleus, with radiation of energy.

Red Shift The shift of spectral lines towards longer wavelengths, caused by the Doppler Effect for a receding source. In cosmology, it refers to the observed shift of spectral lines of distant astronomical bodies towards long wavelengths.

RNA Ribonucleic acid; molecule containing nucleic bases (adenine, guanine, cytosine and uracil) which is involved in protein synthesis.

Satellite A natural satellite is a moon. Artificial satellites now orbit the Earth, taking photographs, sensing variations in heat, light etc on the Earth's surface. Others make environmental measurements, transmit messages and perform other services.

Sediment A comprehensive term for all materials deposited by water, wind or ice. Includes pebbles, silt, mud, sand and some, but not all, clay.

Sedimentary rock Rock formed from accumulations of sediments, which may consist of rock fragments of various sizes and fossil remains.

Shale A fine-grained, sedimentary rock which splits easily. It is made up of clay and silt-sized particles.

Solar System Generally, a sun with a group of celestial bodies held by the sun's gravitational attraction and revolving around it. Specifically, it is applied to the system in which the Earth is one of the circulating bodies.

Speed of light The fundamental constant of special relativity. Equal to 186,000 miles per second.

Steady-State Theory The cosmological theory proposed by Bondi, Gold and Hoyle, in which the average properties of

INDEX

the Universe never change with time; new matter must be continually created to keep the density constant as the Universe expands.

Stromatolite Biogenic sedimentary structures produced as calcium carbonate is deposited by successive mats of algae.

Supernova Enormous stellar explosion in which all but the inner core of a star is blown off into interstellar space. A supernova produces in a few days as much energy as the Sun radiates in about a thousand million years.

Tectonic plate One of the large, rigid or relatively rigid, mobile fragments of the lithosphere at whose boundaries various geological events occur.

Time 'Absolute' time is geological time measured in years. 'Relative' time is the dating of geological events by means of their position in a chronological order of occurrence rather than in terms of years.

Ultra-violet radiation Electromagnetic waves with wavelength intermediate between visible light and X-rays.

Viruses Particles made up of one or several molecules of DNA or RNA and usually coated with protein. They are not living organisms, and can only reproduce within the cells of animals and plants they invade.

Weathering The response of materials that were once in equilibrium within the Earth's crust to new conditions in contact with water, air and living matter.